SPLITTING THE SECOND

The Story of Atomic Time

Tony Jones

INSTITUTE OF PHYSICS PUBLISHING
BRISTOL AND PHILADELPHIA

British Library Cataloguing-in-Publication Data

A catalogue record for this book is available from the British Library.

ISBN 0 7503 0640 8 pbk

Library of Congress Cataloging-in-Publication Data are available

Reprinted 2001

Publisher: Nicki Dennis
Production Editor: Simon Laurenson
Production Control: Sarah Plenty
Cover Design: Victoria Le Billon
Marketing Executive: Colin Fenton

Published by Institute of Physics Publishing, wholly owned by The Institute of Physics, London

Institute of Physics Publishing, Dirac House, Temple Back, Bristol BS1 6BE, UK

US Office: Institute of Physics Publishing, The Public Ledger Building, Suite 1035, 150 South Independence Mall West, Philadelphia, PA 19106, USA

Typeset in TeX using the IOP Bookmaker Macros
Printed in the UK by MPG Books Ltd, Bodmin

Contents

We wish to acknowledge the following for permission to reproduce figures. Science Museum, London (Figure 1.6). Bureau International des Poids et Mesures (Figure 1.7). National Physical Laboratory © Crown Copyright 2000. Reproduced by permission of the Controller of HMSO (Figures 1.8, 2.3, 2.12, 4.6, 4.9, 5.7, 6.5, 6.6, 7.1, 7.4, 7.7, 8.2, 8.4, 8.6). National Institute of Standards and Technology (Figures 2.10, 4.2, 8.5). United States Naval Observatory (Figure 3.2). Physikalisch-Technische Bundesanstalt (Figure 4.1). Bureau National de Metrologie, Laboratoire Primaire du Temps et des Fréquences (Figure 4.5). National Maritime Museum (Figures 5.1, 6.4). National Aeronautics and Space Administration (Figure 5.2). Alan Pedlar and Tom Muxlow, Jodrell Bank Observatory, University of Manchester (Figure 7.5). Figure 8.1 courtesy of the Long Now Foundation.

Foreword

Just fifty years ago, the global time standard was still based on the rotation of the earth on its axis. It was the oldest physical standard in use and also the most accurate. However, in 1955, the National Physical Laboratory developed a new and more accurate time standard, using caesium atoms to set the rate of the clock. Since then, through the efforts of many exceptional individuals and institutions around the world, the atomic clock has transformed the way we measure and use time.

The caesium atom now underpins the very definition of time. The atomic clocks themselves have improved by a factor of nearly a million, with the latest generation using laser-cooled atoms to extract such tremendous accuracy. At this level, Einstein's theory of relativity has become just an everyday engineering tool for comparing the time of atomic clocks. And yet in spite of this extraordinary progress, those at the cutting edge are seeking to exploit alternative atoms to push back the frontiers of time measurement even further.

However, the story told in this excellent book is not just one of scientists breaking through arbitrary boundaries. It is one which affects all our lives. Ultimately we set the time on our watches to a standard maintained by atomic clocks. Telephone networks, electricity grids and satellite navigation systems make full use of the accuracy offered by this technology, and there are countless other examples linking the most advanced and the most mundane of human activities to the beat of the caesium atom.

In spite of its wide spread influence, the story of atomic timekeeping is one that is largely unknown outside a small community of specialists. *Splitting the Second: The Story of Atomic Time* brings up-to-date the traditional account of how we measure and use time. I hope the reader will enjoy this fascinating story.

John Laverty
Head of Time Metrology
National Physical Laboratory
June 2000

Preface

On the wall in my study I have a radio-controlled clock. It is essentially a common-or-garden quartz-crystal clock connected to a tiny radio receiver. Every two hours it tunes in to the rhythmic pulses from a radio station controlled by the atomic clocks at the National Physical Laboratory and corrects itself to Coordinated Universal Time (which—you will soon discover—is commonly, though incorrectly, called Greenwich Mean Time). It adjusts automatically to the beginning and end of summer time and it can even cope with leap seconds, though not in the most elegant fashion. It means we no longer need to wait for radio time signals or to phone the Speaking Clock to get accurate time. It is nice to have a clock guaranteed to remain correct to a tiny fraction of a second, though it is a bit excessive for domestic purposes.

The fact that such clocks and the accuracy they bring are now commonplace is a sign of the upheaval in timekeeping that took place during the twentieth century. It could even be called a revolution. When the century began, timekeeping was firmly in the hands of astronomers, where it had rested for millennia. By the century's end timekeeping was controlled by physicists, and astronomers were relegated to a supporting but not insignificant role. If we were to place dates on the revolution we could say it began in 1955, with the operation of the world's first successful atomic clock, and was all but complete by 1967 when the atomic second finally ousted the astronomical second as the international unit of time.

The start of a new century seems an opportune moment to tell this story, coinciding as it does with the centenary of the National Physical Laboratory. NPL played a central role in that revolution, as you will see, and by a kind of right of conquest is now the official supplier of time to the United Kingdom. Indeed this book owes its origins to Fiona Williams, of NPL, who saw the need for it and has generously supported the project over the past year. I am also grateful to the NPL scientists who have given freely of their time, knowledge and experience, especially John Laverty, James "Mac" Steele, Peter Whibberley and Paul Taylor, and the staff of other institutions who have supplied me with background material and illustrations and answered many queries. I must also thank the staff of the NPL library for their hospitality, Terry Christien for drawing the diagrams and Margaret O'Gorman, Robin Rees and Nicki Dennis at Institute of Physics Publishing who brought the book to fruition.

Tony Jones
May 2000

1

ASTRONOMERS' TIME

A Nobel undertaking

I expect you are reading this book because you are interested in time-keeping. This book is indeed about timekeeping but perhaps not as you have known it. You will find nothing in these pages about balance wheels and verge escapements, nor about the development of the clepsydra or the hemicyclium. And if you wish to know the difference between a foliot and a fusee you will have to look elsewhere.

For this book is about modern timekeeping which, as we shall see, began in June 1955 with the operation of the first atomic clock. The fundamental physics that made the atomic clock possible engaged the minds of many scientists of the first order, and to illustrate that I would like you to look at Table 1.1. Here I have identified 13 winners of the Nobel Prize in Physics since the 1940s. Nobel Prizes are not awarded lightly. Each of these scientists has been honoured for their exceptional work in advancing our knowledge of physics. What they have in common is that all 13 made significant contributions to the science of atomic timekeeping.

Of these only one, Otto Stern, was not concerned with the development of atomic clocks. The rest, from Isidor Rabi onwards, were either working to construct or improve atomic clocks or were conscious of the potential of their work for the accurate measurement of time and frequency.

We shall meet some of these laureates in the book, though only briefly, for this is not primarily a history of the atomic clock but an account of timekeeping today. To gain a perspective on the revolution that the atomic clock has brought in its wake we shall nonetheless have to look at some history, and we shall start with the oldest method of timekeeping—the Sun.

Table 1.1. Some Nobel Laureates in physics.

Year of award	Nobel Laureate	Contribution to atomic timekeeping
1943	Otto Stern	Stern showed how beams of atoms could be used to investigate the magnetic properties of atoms and nuclei
1944	Isidor Rabi	Rabi, who had worked with Stern for two years, developed the "atomic beam resonance method" for investigating the magnetic properties of nuclei. He was the first to propose that a beam of caesium atoms could be used to make an atomic clock
1955	Polykarp Kusch	Kusch, a colleague of Rabi, was one of the experimental pioneers of atomic clocks. His practical design inspired the construction of the first operational atomic clock at the National Physical Laboratory
1964	Nikolai Basov, Aleksander Prochorov, Charles Townes	These physicists independently invented the type of radiation amplifier known as a maser or laser; the maser would open the way to a second type of atomic clock. Townes was a former colleague of Rabi
1966	Alfred Kastler	Kastler invented the technique of "optical pumping" which is now used in the most sensitive caesium clocks
1989	Norman Ramsey	A former colleague of Rabi, Ramsey made two quite different contributions. He devised the "Ramsey cavity", an essential component of all caesium clocks, and went on to build the first hydrogen maser clock
1989	Hans Dehmelt, Wolfgang Paul	Dehmelt and Paul invented methods of isolating and trapping single atoms which are now being used in fundamental research into the atomic clocks of the future
1997	Steven Chu, Claude Cohen-Tannoudji, William Phillips	These three devised methods for cooling atoms to within a fraction of a degree of absolute zero. Their techniques are vital to the latest types of atomic clocks, the caesium fountains

Solar time

For practically the whole of human history, up to the latter decades of the twentieth century in fact, our timekeeping has been based on the apparent motion of the Sun across the sky. Apparent, because it is the rotation of the Earth on its axis that sweeps the Sun across the sky every 24 hours rather than any movement of the Sun itself. In using the Sun to define our scale of time, we are relying on the unceasing spin of the Earth to count out the days.

How long is a day?

Imagine a great semicircle drawn on the sky from the north point on the horizon, through the zenith (the point immediately above your head) and down to the south point on the horizon (Figure 1.1). This line is called the meridian and it divides the bowl of the sky into an eastern half and a western half. Now we can define the length of the day more precisely. When the Sun crosses the meridian it is noon. The time between two successive meridian crossings we shall call a "day". Note that this definition is unaffected by the need to see the horizon—it doesn't matter when the Sun rises or sets. Neither is it affected by the varying length of daylight through the year. The Sun's crossing of the meridian gives us both the instant of noon and the duration of the day—it defines both a time scale and a unit.

It comes as a surprise to many people that the length of the day defined in this way varies through the year. If we were to time successive meridian crossings with an accurate clock we would find that the length of the day kept by the Sun varies from 22 seconds short of 24 hours (in September) to 30 seconds in excess (in December) and it rarely crosses the meridian precisely at 12 o'clock. What's going on?

To understand this we need to look more closely at the motion of the Sun. As the Earth completes a single orbit of the Sun each year, the Sun appears to us to make a corresponding circuit about the Earth in the same time. The path of the Sun around the sky is called the ecliptic. If we could see the background stars we would notice the Sun creeping eastwards along the ecliptic at about one degree every day (because a complete circle is 360 degrees and there are 365 days in the year). To be

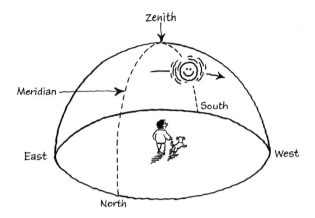

Figure 1.1. "Noon" is defined as the moment the Sun crosses the meridian, an imaginary line extending from the north to the south horizons and passing through the zenith. The solar day is the interval between successive noons.

precise, if the Earth's orbit were circular the speed of the Sun around the ecliptic would be an unchanging 0.986 degrees per day.

But like virtually all astronomical orbits, the Earth's path is an ellipse, and this is the first reason for the changing length of day. The Earth is a full 5 million kilometres closer to the Sun on 3 January than it is on 4 July, give or take a day either way. At its nearest point to the Sun, the Earth is moving faster in its orbit than at its furthest point. Seen from the Earth, the Sun appears to skim along at a brisk 1.019 degrees a day in January, while at the height of summer it moves at a leisurely 0.953 degrees a day. By itself, this effect would give us shorter days in the summer than in the winter.

A second reason why the length of the day is not constant is that the Earth's axis is tilted with respect to the plane of its orbit, which means that the ecliptic is inclined to the equator by the same amount. This is why the Sun appears to move northwards in the spring and southwards in the autumn. Only at the solstices, near 21 June and 21 December, is the Sun moving directly west to east; at all other times some part of the Sun's motion is directed either north or south and it does not progress

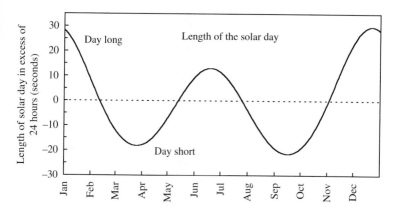

Figure 1.2. Because the Earth's orbit is not circular and the Earth's axis is tilted, the length of the solar day varies through the year. It is almost a minute longer in late December than in mid-September.

so fast around the sky. By itself, this effect would give us longer days in summer and winter and shorter days in spring and autumn.

Taken together, these two effects cause the length of the day to vary in the complex manner shown in Figure 1.2. Makers of sundials have always known this, and many ingenious methods have been devised to make the dials read the right time. But a day that varies through the year is not much use for precise timekeeping, so astronomers introduced the notion of the "mean sun", an imaginary body that moves steadily around the equator—rather than the ecliptic—at a precise and uniform speed. The concept of the mean sun is just a mathematical way of straightening out the effects of the elliptical orbit and the tilt of the Earth's axis to create a "mean solar day" that is always the same length. The time kept by the mean sun is known as mean solar time, while the time kept by the real Sun (and shown on a sundial) is apparent solar time. They can differ by more than 16 minutes, a discrepancy known as the "equation of time" (Figure 1.3). The true Sun and the mean sun both return to the same position after exactly one year, so in the long run mean solar time keeps step with apparent solar time.

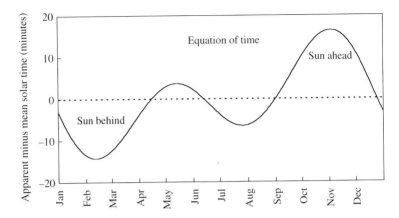

Figure 1.3. The "equation of time" is the difference between apparent and mean solar time due to the changing length of the solar day. The Sun is more than 14 minutes behind mean solar time in mid-February and more than 16 minutes ahead in early November. A sundial only shows mean solar time at four dates in the year: near 16 April, 14 June, 2 September and 25 December. If you want to set your sundial to read as close as possible to the correct time, these are the dates to do it.

Mean solar time was the basis for all timekeeping until the last few decades. Apparent solar time still has its uses, especially in traditional navigation at sea. Indeed, the US Nautical Almanac continued to use apparent solar time in its tables as late as 1833.

Standard time

An obvious drawback of timekeeping based on the Sun—even the mean sun—is that it varies around the world. If noon is defined as the moment when the mean sun is on the meridian, then solar time will be different at different longitudes. Noon in London comes about 10 minutes after noon in Paris and 54 minutes after noon in Berlin. Yet it comes 25 minutes before noon in Dublin and almost 5 hours before noon in New York. If you happened to live on Taveuni Island in Fiji—at longitude 180 degrees—noon in London would coincide precisely with local midnight, which is why the Fijians were able to greet the millennium a full 12 hours before Londoners.

Until the last century everyone lived quite happily with their own local version of mean solar time. When the pace of life was slower and people didn't travel very fast it didn't matter that the time in Manchester was 3 minutes ahead of that in Liverpool, or even that clocks across North America could differ by several hours. But with the coming of the telegraph and the railways, there was a pressing need to agree on what the time was across distances of hundreds or thousands of kilometres. How could trains run on time if no one agreed what the right time was?

The solution—first introduced in the US and Canada in 1883—was to divide up the country into "time zones". In each zone the clocks would all read the same, and clocks in neighbouring zones would differ by precisely 1 hour. The idea caught on and in 1884 an international conference in Washington set up a system of time zones for the whole world. The basis of world time would be mean solar time at the Royal Observatory at Greenwich, in east London, which from 1880 had became legally known as Greenwich Mean Time, or GMT. (In fact, GMT no longer exists but we'll use the term in this chapter until the full truth can be revealed ...)

In theory, time zones divide up the world into 24 zones of 15 degrees in longitude—rather like segments of an orange. Each zone has its own standard time, based on mean solar time at the central longitude of the zone, and differing by multiples of 1 hour from GMT. Everywhere between longitude $7\frac{1}{2}$ degrees east and $7\frac{1}{2}$ degrees west is within the Greenwich time zone and clocks read GMT. Between $7\frac{1}{2}$ and $22\frac{1}{2}$ degrees west clocks read GMT minus 1 hour, and between $7\frac{1}{2}$ and $22\frac{1}{2}$ degrees east clocks read GMT plus 1 hour. In this way the world can be divided up into 15-degree segments east and west of Greenwich, until we get to the other side of the world. The time zone exactly opposite to Greenwich is centred on longitude 180 degrees and differs by 12 hours, but is the standard time there 12 hours ahead or 12 hours behind GMT? The answer is both; the zone is split down the middle by the International Date Line. On either side of the Date Line the standard time is the same, but the date differs by one day.

In practice, the world's time zones have been heavily influenced by geography and politics and bear little resemblance to their theoretical boundaries. Even the Date Line has a few kinks in it to avoid populated areas. It is up to each country to decide which time zone it wishes to adopt. Most of Western Europe is on Central European Time (GMT + 1 hour), even countries like France and Spain which according to their longitude should be on GMT. In these countries noon occurs nearer to 13:00 mean solar time than 12:00. China covers three time zones, but all the clocks are set to 8 hours ahead of GMT. In a few places the zone time differs by fractions of an hour from GMT; Newfoundland is $3\frac{1}{2}$ hours behind GMT while Nepal is $5\frac{3}{4}$ hours ahead. Areas near the poles, like Antarctica, have no standard time at all and use GMT instead. Inconsistent it may be, but what matters is that the standard time at every point on Earth has a known and fixed relationship to GMT.

Universal time

In 1912 the French Bureau des Longitudes convened a scientific conference to consider how timekeeping could be coordinated worldwide. The conference called for an international organisation to oversee world timekeeping. The following year a 32-nation diplomatic convention established an Association Internationale de l'Heure intended to supervise a Bureau International de l'Heure (BIH) which would carry out the necessary practical work. A provisional bureau was set up at once, but with the outbreak of World War I the convention was never ratified and the infant BIH, based at Paris Observatory, continued as an orphan until it was taken under the wing of the newly formed International Astronomical Union (IAU) in 1920. One of the major activities of the BIH was to correlate astronomical observations to create a worldwide system of timekeeping.

One early problem to be tackled concerned the definition of GMT itself. Astronomers tended to work at night, and it was a nuisance for the date to change midway through their working day, at least for those in Greenwich (astronomers in Fiji would have been quite happy). So astronomers had always reckoned GMT from noon to noon rather than midnight to midnight. (Astronomers were not uniquely perverse: until well into the nineteenth century the nautical day was also reckoned from noon to noon, but what the astronomers called Monday the sailors called Tuesday ...)

This confusing state of affairs, with astronomers being 12 hours behind everyone else, lasted until 1925 when the IAU redefined GMT so that it always began at midnight, even for astronomers. So 31 December 1924 was abruptly cut short, with 1 January starting only 12 hours after 31 December. Astronomers' GMT beginning at noon was redesignated Greenwich Mean Astronomical Time (GMAT). Yet the confusion persisted and in 1928 the IAU replaced GMT with a new designation, Universal Time (UT). UT is the mean solar time on the Greenwich meridian, beginning at midnight.

So for the first time the world had a clear and unambiguous time scale that everyone agreed on. Universal Time was based on the mean solar day which was determined from astronomical observations. The day was divided into 86 400 seconds; thus the scientific unit of time, the second, was tied to the rotation of the Earth.

Summer time

We should mention one more variant on mean solar time. Many countries like to "put the clocks forward" in the spring to give people an extra hour of daylight on summer evenings. The 15 countries of the European Union, for example, advance all their clocks by 1 hour at 01:00 GMT on the last Sunday in March and put them back by 1 hour on the last Sunday in October.

When summer time (or "daylight saving" time) is in force the Sun rises an hour later according to the clock, crosses the meridian an hour later and sets an hour later than it otherwise would. (In countries like Spain, which are normally 1 hour ahead of their zone time anyway, this means that noon occurs at about 14:00.) Of course this has no effect whatever on the actual hours of daylight, it just gives the illusion of longer evenings. What actually happens is that everyone gets up an hour earlier than they would otherwise do. If the government told everyone to get up an hour earlier in the summer there would be a public outcry, but that is precisely what happens under the guise of "summer time".

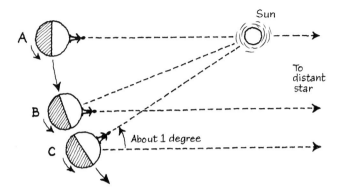

Figure 1.4. The sidereal day is slightly shorter than the solar day. At point A the star is on the meridian at the same time as the Sun. When the Earth has rotated to B the star is once again on the meridian—a sidereal day has passed—but the Earth has to turn through a further small angle before the Sun returns to the meridian and a solar day has passed. The difference is about four minutes of time or one degree of angle.

Sidereal time

We have said that UT is determined by astronomical observation. Although based on the mean solar day, UT has never been reckoned by measurements of the Sun, except by navigators at sea. On the sky the Sun is half a degree wide. It takes 2 minutes to move through its own diameter, so it is actually very difficult to measure the position of this blazing disk of light with great accuracy. And the mean sun, being imaginary, is not observable at all.

In practice, astronomers measure time by observing the stars. Like the Sun, the stars rise and set and move across the sky. By observing stars crossing the meridian, rather than the Sun, astronomers defined a sidereal day. But there is a subtlety. The time between two successive crossings of the meridian by a star, a sidereal day, is slightly shorter than a mean solar day. To be precise, it is 23 hours, 56 minutes and 4 seconds.

To see why this is, look at Figure 1.4. At position A the Sun and a star are both on the meridian (though the star would not be visible in daylight of course). At position B, a day later, the Earth has made a full turn so that the star is back on the meridian. But now the motion of the Earth has carried it some way around its orbit and the Sun has not yet reached the meridian. The Earth has to turn a little further—about one degree—before the Sun crosses the meridian and a solar day has passed. This further turn takes 3 minutes and 56 seconds, and over the course of a year adds up to an extra day. So a year is made up of 365 solar days but 366 sidereal days.

Because the mean sun moves at a steady and fixed rate with respect to the stars, the relationship between the lengths of the sidereal day and the mean solar day is also fixed. So UT was measured by first timing the transits of stars to find sidereal time and then applying a correction to obtain Universal Time.

Just as solar time tells us the orientation of the Earth with respect to the Sun, sidereal time is a measure of the orientation of the Earth with respect to the stars. Every astronomical observatory has a clock set to show local sidereal time (LST). At about 17:46 LST, for example, astronomers know that the centre of the Galaxy is on the meridian and so is best placed for observation. If they want to observe the Orion Nebula, it is on the meridian at 05:35. The Andromeda Galaxy is at its highest in the sky at 00:43. Sidereal time coincides with mean solar time at the spring equinox and then runs fast at a rate of about four minutes a day until a complete day has been gained by the following spring.

Sidereal time is measured in hours, minutes and seconds, each of which is slightly shorter than the mean solar hour, minute and second. Like solar time, sidereal time is different at each longitude, and astronomers use a Greenwich Sidereal Time which is analogous to Greenwich Mean Time.

Something wrong with the Earth

By the 1920s astronomers had a supposedly uniform time scale, Universal Time, that was based on the mean motion of the Sun, which of course reflected the rotation of the Earth, but was measured by timing the apparent motion of the stars. UT was adopted worldwide, both for scientific and civil timekeeping. Yet long before then there were inklings that all was not well with the rotation of the Earth.

Precession

Even in the second century BC, the Greek astronomer Hipparchus had discovered that the Earth's axis is not fixed in space. Like a spinning top, it slowly traces out a circle on the sky once every 25 800 years. At the moment the north pole points very nearly towards Polaris (which takes its name from being the pole star), but 4500 years ago it pointed roughly to Thuban in the constellation of Draco and around the year 14 000 it will be near the bright star Vega. Imposed on this circular motion is a slight wobble called nutation. Precession and nutation are caused by the gravitational tug of the Sun and Moon on the Earth's equatorial bulge, but the effects are predictable and can be allowed for.

The lengthening day

Early indications that something was wrong with the Earth's rotation came from observations of the Moon. In the seventeenth and eighteenth centuries many astronomers were concerned with the problem of finding longitude at sea, which was really a question of timekeeping. Though the answer would ultimately come from an improved chronometer rather than from astronomy, one promising idea was to use the Moon as a kind of celestial clock. Just as the hands of a clock sweep over its face, the Moon sweeps around the sky once a month. If the movements of the Moon could be predicted accurately, a navigator could measure the position of the Moon against neighbouring stars and look up the time in a table.

In 1695 Edmond Halley, one of the more accomplished scientists of the time, published a study of ancient eclipses. He had examined records of eclipses to work out the position of the Moon in the distant past, but could not reconcile the ancient observations with modern ones. The only way he could make sense of them was if the Moon were now moving faster in its orbit than it was in the past.

This notion was confirmed in 1749 by Richard Dunthorne, who used the ancient eclipse observations to calculate that the Moon had drifted ahead of its expected position by almost two degrees over a period of more than 2400 years. How such an acceleration could be produced was investigated by the leading mathematicians of the time, but they could not make the Moon speed up.

A solution appeared to come in 1787, when French mathematician Pierre-Simon Laplace proposed that the movements of the planets distorted the shape of the Earth's orbit. This in turn affected the pull of the Sun on the Moon which led to the Moon's steady acceleration. Laplace's calculations were in good agreement with the findings of Dunthorne and others and the discovery was regarded as a crowning achievement of celestial mechanics. However, in 1853 British astronomer John Couch Adams, who had successfully predicted the existence of Neptune a few years earlier, repeated the calculations to higher precision and showed that Laplace's theory accounted for only half of the Moon's acceleration, but his result was not widely accepted.

Tidal friction

It was not until the 1860s that it finally dawned on astronomers that at least part of the apparent acceleration of the Moon could be due to a *deceleration* of the Earth. If the Earth's rotation were gradually slowing, the mean solar day would no longer be constant but lengthening. And with it would lengthen the hour, the minute and the second. If the units of time were lengthening, what would be the effect on the Moon?

Suppose that the motion of the Moon around the Earth were uniform. That is to say, in any fixed interval of time the Moon moves through precisely the same arc in its orbit around the Earth. If the Earth were slowing down, causing the day to lengthen, the Moon would appear to move very slightly further each day than the previous day. If we didn't know about the slowing of the Earth we would see the daily motion of the Moon appear to increase—to our eyes the Moon would appear to be accelerating. Over many centuries the discrepancy between where the Moon ought to be and where it actually is would become appreciable. This is what Halley and his successors were grappling with when they tried to reconcile ancient and modern observations.

But how could the Earth be slowing down? The answer came, independently, from US meteorologist William Ferrel and French astronomer Charles-Eugène Delaunay, and it was to do with the Earth's tides. The twice daily rising and falling of the tides are familiar to everyone. They are caused, of course, by the gravitational pulls of the Moon and, to a lesser degree, of the Sun. The gravitational attraction of the Moon falls

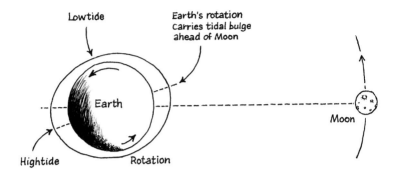

Figure 1.5. The Moon raises two tidal bulges in the Earth's oceans, which are carried ahead of the Moon by the Earth's rotation. Friction between the raised water and the sea bed dissipates energy at the rate of 4 million megawatts, and slows the rotation of the Earth. At the same time the Moon is gradually pushed away from the Earth.

off with distance. It follows that the attraction on the near side of the Earth is slightly greater than the attraction on the far side. The result is a net stretching force that tends to pull the Earth into a rugby-ball shape in the direction of the Moon. Because water can flow more readily than the solid body of the Earth, the oceans heap up into two bulges about half a metre in height, one facing the Moon and one on the opposite side. As the solid Earth turns beneath the bulges, we see the oceans rise and fall (see Figure 1.5).

The Earth rotates faster than the Moon revolves around it, and so the tidal bulges are carried slightly ahead of where they would be if the Earth were not rotating. This is why high tides occur an hour or so before the Moon crosses the meridian. But this dragging of the bulges has a cost in terms of friction between the oceans and the ocean bed, especially in the shallow zones around the continental shelves.

Ferrel and Delaunay showed that the frictional heating caused by the tides, amounting to some 4000 billion watts, would result in a measurable slowing of the Earth's rotation. The bulges are acting like the brake shoes on the wheel of a car, gradually slowing the Earth and turning its rotational energy into heat. In other words, the day is becoming longer because of the tidal drag.

Another consequence of tidal drag is the loss of angular momentum. One of the principles of physics is that angular momentum cannot be created or destroyed. If the Earth is losing angular momentum as it slows, then it must be going somewhere else. Where to? Ferrel and Delaunay showed that it is being transferred to the Moon. The Moon is gaining angular momentum and it is terribly easy to leap to the conclusion that the Moon is speeding up as the Earth slows down and that this is the observed "acceleration" of the Moon. But, no, it's not that straightforward. Simple physics shows that as the Earth slows down the Moon moves further away from us at about 3 or 4 centimetres a year. As it drifts away the Moon moves more slowly in its orbit. So the slowing of the Earth's rotation actually causes a *deceleration* of the Moon in its motion around the Earth; only if our measure of time is locked to the lengthening mean solar day does this appear as an acceleration. No wonder astronomers were confused.

Tidal drag works both ways. Though the Moon has no oceans, the much stronger gravity of the Earth raises tides in the solid body of the Moon. The deformation is about 20 metres and the creaking of the Moon can be detected as "moonquakes" with seismic instruments left by the Apollo astronauts. In fact, tidal drag on the Moon has stopped the rotation completely, which is why it keeps the same face towards the Earth. One day the Earth's rotation will stop too, and the Moon will appear to hang motionless in the sky above one hemisphere of the Earth and be forever hidden from the other. Perhaps travel companies will do a brisk trade in tours from the moonless side of the Earth to the moonlit hemisphere.

But the steady slowing of the Earth by tidal drag could not be the whole story. From the mid-1800s observations of the Moon showed that its "acceleration" was not the steady change predicted from tidal drag. Even with tidal effects allowed for, the Moon was sometimes ahead and sometimes behind its expected position, and the changes took place on time scales of decades. Yet, despite the discovery that the Earth was slowing, astronomers were reluctant to concede that these irregular variations might stem from fluctuations in the Earth's rotation rather from the dynamics of the Moon. By 1915 all alternative explanations—

invisible matter in the Solar System, magnetic forces and even swarms of meteorites—seemed to have been exhausted.

Of course, if the Earth's rotation really were unsteady then not only the Moon but also the Sun and all the other planets would show similar fluctuations. These discrepancies were much harder to detect since all of these bodies move more slowly in their paths than does the Moon. It was not until 1939 that Sir Harold Spencer Jones, the Astronomer Royal for England, showed conclusively that the Moon, Sun and Mercury had been displaying proportionately the same discrepancies since reliable telescopic observations became available in the late seventeenth century. It was not the movements of the celestial bodies that were fluctuating, but the rotation of the Earth and with it the units of time.

Chandler's wobble

In the 1880s came yet another discovery, though one that had been suspected for some time. Very precise measurements of the positions of stars through the year showed that the latitude of astronomical observatories was changing by a tiny amount. This meant that the positions of the Earth's poles were moving. Seth Chandler, Harvard astronomer and former actuary, analysed observations going back 200 years and announced that there were two sets of motions: an annual motion and a motion with a period of 428 days. This "polar wobble" is not to be confused with the motion of the poles during precession—it is not a matter of the direction of the polar axis turning in space, but the axis itself is moving over the ground. If the north pole could be represented by a physical post, we could stand on the ice and see it tracing out a rough circle several metres in diameter each year.

Seasonal variations

By the 1930s scientists in France and Germany, using the latest accurate clocks, were finding still another problem with the Earth's rotation. It now appeared that the length of the day depended on the time of year. This was something like a miniature version of the "equation of time", with the Earth running as much as 30 milliseconds late in spring and a similar amount ahead in the autumn.

So by the 1940s it was clear that not only was the day steadily lengthening, and with it the hour, minute and second, but the lengthening was not uniform. The day was shorter in summer than in winter, if only by a millisecond or so, the poles were wobbling and, worst of all, there were seemingly irregular fluctuations that were perhaps rooted in unknown and unknowable processes occurring deep inside the Earth. Many astronomers had come to the uncomfortable conclusion that they could no longer depend on the Earth as the world's timekeeper.

Ephemeris Time

If the rotation of the Earth could no longer be relied on to provide a uniform time scale, what was the alternative? An early proposal came from André Danjon of the University Observatory at Strasbourg. In an article in 1927, Danjon proposed that astronomers abandon time reckoning based on the rotation of the Earth and instead develop an alternative based on the motions of the planets in their orbits around the Sun. In essence he was proposing that the basis of timekeeping should be the year rather than the day.

This made a lot of sense. Ever since Isaac Newton showed how the planets moved in accordance with a single universal law of gravitation, the notion of the Solar System as being like a majestic system of clockwork had had wide appeal. In many science museums you can see a mechanical model of the Solar System called an orrery (see Figure 1.6). The model planets move in their orbits at the correct relative speeds, driven through a system of interlocking gears. In reality the planets move independently—there are no gears—but such is the uniformity of physical law that the Solar System does behave as if the orbits are locked together, driven by a hidden motor whose steady turning controls the movements of all the planets.

The regular beat of time which guides the planets has been called Newtonian time. This is the time which astronomers used to predict the positions of the planets at regular intervals into the future. By definition, Newtonian time flows smoothly, without the irregularities of the rotating Earth.

Little happened as a result of Danjon's idea until 1948 when Gerald

Figure 1.6. An example of an orrery from approximately 1800. The Sun is the large ball in the centre while the planets, from left to right, are Uranus, Saturn, Jupiter, Venus, Mercury, Earth and Mars. Their moons are also shown. Turning the handle (right) works a system of gears which moves the planets and their moons at the correct relative speeds. The rate at which the handle is turned is analogous to Ephemeris Time.

Clemence, of the US Naval Observatory, published a detailed proposal for a very similar system. Clemence proposed that the time used by astronomers to calculate the position of the Sun should become the new basis of timekeeping. Since the turn of the century the position of the Sun had been calculated from a formula devised by Simon Newcomb, an astronomer at the US Nautical Almanac Office. Newcomb's formula gave the position of the Sun for any desired time. For practical purposes the calculated positions were listed in a table, known as an ephemeris: you look up the date and time in the ephemeris and out comes the position of the Sun. Clemence showed how the ephemerides of the Moon and planets could be modified so that they all used the same Newtonian time as the ephemeris of the Sun.

Reading the time, in principle, was then straightforward. No longer would time be measured by observing the passage of stars across the meridian. Instead, you measure the positions of the Moon and planets against the stars, and look up in the ephemeris the time at which they are predicted to be in those positions. Because of the interlocking "gears", you can tell the time by looking at the motions of any of the bodies, and the answer should be the same.

In 1950 Clemence presented his ideas to a conference organised by the IAU in Paris convened by Danjon, who had then become Director of the Paris Observatory. The conference recommended that Newcomb's measure of time be adopted. It was to be called Ephemeris Time (ET), a name suggested by Dirk Brouwer, an astronomer at Yale University. The basic unit of ET was to be the length of the sidereal year in 1900, that is, the time taken by the Earth to complete one orbit of the Sun with respect to the stars. It so happens that the length of the year is not constant—that is why the year 1900 was specified—but the changes were small and well understood. The resolution was adopted by the IAU General Assembly in 1952.

"For all people, for all time"

We shall leave the astronomers for a while and take a look at what the rest of the world was doing about measuring time. Until the 1950s few outside the scientific community had thought very hard about units of time. Even the scientific unit of time, the second, was regarded as 1/86 400 of a day with the unstated assumption that a day—being the mean solar day—was a fixed length of time.

This relaxed attitude to time contrasted sharply with the highly systematic definitions of other physical quantities. Moves towards defining a rational set of units of measurement originated in post-revolutionary France in the 1790s. The French Academy of Sciences was charged with setting up a system of units to replace the multitude of customary measures then in use in France.

They started from the principle—which has also guided their

successors—that units should not be arbitrary, like the volume of a
barrel or the length of the king's forearm, but should have some basis
in nature. First came a new unit of length, the metre. When conceived
in the 1790s, the metre was defined as one-ten millionth of the distance
from the north pole to the equator, measured along the meridian of Paris.
Surveyors spent six hazardous years measuring the meridian through
France and Spain at a time when the two countries were on the brink of
war, and the result of their labours was preserved in the form of a bar of
platinum whose length was declared the legal metre in 1799. Next came
the unit of mass. The gram, originally defined to be the mass of a cubic
centimetre of water at four degrees Celsius, was realised in the shape
of a 1000-gram platinum cylinder, the kilogram. The founders of the
metric system expressed the hope that it would in time form the basis
for international agreement on a system of units "for all people, for all
time".

Attempts to rationalise the measurement of time were not so successful. The Academy proposed that the day be divided into 10 new
hours, each of 100 new minutes, each of which comprised 100 new
seconds. That would have meant 100 000 new seconds in the day, each
new second measuring 0.864 mean solar seconds. But there was still no
question over the basic unit of time, the mean solar day. The proposal
was abandoned in 1795 in the face of stiff resistance, but not before a
few 10-hour clocks had been built.

Moves towards world agreement on units of measurement began in
1875, with the signing in Paris of the Convention du Mètre (the Metre
Convention) by 17 nations. The convention set up the Bureau International des Poids et Mesures (BIPM; International Bureau of Weights
and Measures), whose job it was to administer the new standards (we
shall hear a lot more about BIPM later in this book). It was (and still
is) supervised by the Comité International des Poids et Mesures (CIPM)
which was in turn accountable to the Conférence Générale des Poids
et Mesures (CGPM), made up of delegates from member governments
meeting every four years. The CGPM remains the ultimate authority
for definitions of units—when the CGPM defines a unit, that is what it
is. In recognition of French leadership in promoting the new system of
measurement, BIPM was given a home at the Pavillon de Breteuil in

Figure 1.7. The Pavillion de Breteuil, the headquarters at Sèvres, near Paris, of the Bureau International des Poids et Mesures (BIPM), the keeper of the world's standards of measurement. The building and the surrounding grounds have the legal status of an embassy.

Sèvres, on the outskirts of Paris, where it remains to this day (Figure 1.7).

The first meeting of the CGPM in 1899 saw the unveiling of the new International Metre and International Kilogram, each based as nearly as possible on the French standards of a century earlier. The International Metre was a bar of platinum–iridium alloy kept at BIPM. The metre was, by definition, the distance between two fine scratches on the bar when measured under certain conditions. Twenty-nine copies of the bar were distributed to national standards laboratories and they would be periodically taken back to Sèvres to check that they were still accurate. The British copy, for example, is kept at the National Physical Laboratory (now relegated to the museum) and had been recalibrated at BIPM on six occasions by the 1950s.

The International Kilogram, which remains the world standard for mass, was a solid cylinder of platinum–iridium alloy (the same material as the International Metre) also kept at BIPM. The British copy has been recalibrated four times and shows agreement with the prototype to better than one part in 100 million (see Figure 1.8).

Figure 1.8. The UK copy of the international prototype kilogram which is kept at the National Physical Laboratory. The second cannot be defined so easily.

But still there was no International Second. Time, of course, is different from length and mass. It is not possible to make a platinum–iridium casting of a second to serve as an international standard. Time is altogether of a different quality, and until the 1950s the second was taken to be 1/86 400 of a mean solar day, on the assumption that the length of the day was easily measurable by observation and, moreover, fixed. And it was the job of the astronomer to deliver the length of the day.

A new second

By the 1950s the CGPM was engaged in a much more ambitious process of rationalising all units of measurement, both commercial and scientific, to form a consistent system that could be applied worldwide. The new

Système International d'Unités (International System of Units, known as the SI) would establish six base units (later seven) upon which all other units of measurement could be constructed. The CIPM followed with interest the debate in astronomy about Ephemeris Time and saw the opportunity to formulate a precise definition of the second. In 1956 it established a committee of representatives from the IAU and national standards laboratories to advise on a definition of the second that could be integrated into the new SI.

Since deciding to adopt ET in 1952, the IAU had revised the proposed definition and now favoured basing Ephemeris Time on the duration of the so-called "tropical" year rather than the sidereal year. The reasoning was that, although the stars provide a sound frame of reference against which to measure a complete orbit of the Earth, that is not the year which actually matters in scientific and everyday life. A more meaningful year is one that keeps pace with the seasons and is measured from one spring equinox to the next. Because of the precession of the Earth's axis, this "tropical" year is a full 20.4 minutes shorter than the sidereal year. If the sidereal year had been chosen, the seasons would have started slipping around at rate of one day every 70 years. By the year 4000 the spring equinox would be occurring in February and the midwinter solstice in November.

After discussions with the IAU, the CIPM decided on a formal definition of the second that would be consistent with the new scale of Ephemeris Time. In 1956 they recommended that the SI second should be the fraction 1/31 556 925.9747 of the tropical year for 1900 January 0 at 12 hours Ephemeris Time (January 0 1900 is just another way of saying 31 December 1899). With the second now defined precisely for the first time, all that remained was to fix the starting point for the new Ephemeris Time. In 1958 the IAU declared that "Ephemeris Time is reckoned from the instant, near the beginning of the calendar year AD 1900, when the geometric mean longitude of the Sun was 279° 41′48.04″, at which instant the measure of Ephemeris Time was 1900 January 0d 12h precisely." These rather cumbersome definitions were chosen with care. The figures were derived from the formula devised by Simon Newcomb for the ephemeris of the Sun, and the definitions ensured that the new ET would mesh smoothly with earlier observations of the Sun. (It did not seem to matter much that Newcomb's formula was based on observations made as long ago as 1750 but we shall later see the consequence of adopting a unit of time rooted so firmly in the past.)

From 1960 Ephemeris Time began to appear in astronomical tables and in the same year the new SI was ratified by the CPGM. The world at last had a coherent system of units including the traditional trio of mass, length and time. The unit of mass was the kilogram, defined as the mass of the platinum–iridium cylinder kept in Sèvres since 1889. Anyone could go to Paris with their national standard kilogram and calibrate it against the world standard. The unit of length was the metre, now defined in terms of the wavelength of light from a krypton lamp. Anyone with suitable equipment could make a metre in their own laboratory. And time? The unit of time was the second, and the second was defined as 1/31 556 925.9747 of the tropical year for January 0 1900, at 12.00 Ephemeris Time. So how, in practice, could you make a second?

2

PHYSICISTS' TIME

Introduction

So far in our discussion of timekeeping something has been missing. We have seen how time scales are based upon the rotation of the Earth or the movements of the Sun, Moon and planets. Yet when it comes to practical matters, we do not glance at the sky to find out if it is time for lunch or to fetch the children from school. That is why we have clocks. The traditional purpose of a clock was to subdivide the mean solar day into the more convenient intervals of hours, minutes and seconds. Every now and then the clock would be adjusted to mean solar time by comparing it with a better clock, which ultimately derived its time from signals emanating from astronomical observatories which computed the time from measurements of the stars.

Used in that way clocks are secondary standards of time that maintain a time scale between periodic calibrations. But why not base a time scale on the clock itself? There is no reason why a sufficiently good clock should not provide a time scale completely independent of the Earth's rotation, the movements of the planets, or any astronomical phenomenon, and do it better. In principle all our timekeeping needs could be met by a black box in the basement of a standards laboratory, ticking out seconds regardless of what is happening in the sky overhead. And to a very great extent that is exactly what happens today. But to see how that has come about, we need first to consider what makes a good clock and how we could make a clock that keeps time better than the Earth.

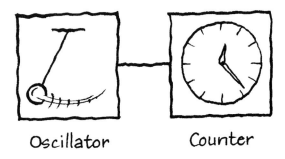

Oscillator   Counter

Figure 2.1. Every clock can be thought of as having two parts: an oscillator which provides a steady beat, and a counter which counts and displays the cycles.

How good is a clock?

Any clock can be thought of as having two parts: an oscillator and a counter (Figure 2.1). The oscillator—or frequency standard—is the part that provides a repetitive, periodic vibration of some kind. It may be a swinging pendulum, a balance wheel, a vibrating crystal, the rhythmic fluctuations of mains alternating voltage, or the vibration of electrons in atoms. It could be the turning of the Earth or the turning of the Solar System. Whatever it is, whether fast or slow, the oscillator supplies the rhythmic, regular beat that drives the clock.

But an oscillator alone is not a clock, for it does not tell the time. Think of a musician's metronome. It ticks off regular beats but it does not count them. It is an oscillator but not a clock. To make an oscillator into a clock we need a mechanism—a counter—to keep a tally of the beats and display the accumulated total. The counter keeps track of how many cycles have passed. It makes the difference between a frequency standard and a real clock. In an old-fashioned "clockwork" watch the oscillator is a balance wheel, a coiled spring that swings to and fro. The counter is the escapement mechanism that ticks off the cycles and moves the hands to indicate the time. In a digital watch the oscillator is a vibrating crystal and the counter is an electronic circuit which shows the time in a numerical display.

If we are to construct a clock that keeps time better than the Earth, we need some way of measuring how good it is. Timekeeping professionals use two measures to describe the goodness of a clock—its accuracy and its stability.

Accuracy is the ability of a clock to read the correct time. A clock may be inaccurate either because its rate is not correct—it runs too fast or too slow—or because it has not been set correctly. The first of these, *frequency accuracy*, is the more fundamental. It is essentially a measure of how well the clock ticks out nominal intervals of time, such as seconds. If the ticks are very close to coming at 1-second intervals, then the clock has a high frequency accuracy. A common measure of accuracy is how much a clock is in error after running for one day. If you set a clock to the correct time and find 24 hours later that it is 1 minute slow, its accuracy is 1 minute a day, or one part in 1440, or about seven parts in 10^4.

Stability is the ability of a clock to run at a constant rate. Or to put it another way, a stable clock adjusted to tick once a second will continue to tick once a second. If the clock losing 1 minute a day continues to lose a minute every day then it is quite a stable clock, even if not very accurate. On the other hand, if it lost 1 minute the first day, 2 minutes the second day and gained a minute the third day it would not be very stable. Stability is measured over a stated interval of time. If a clock measures out 24 hours to within plus or minus 2 minutes each day, then the stability is 2 minutes a day, or about 1.4 parts in 10^3.

It follows that, provided a clock is stable, its accuracy can often be improved by measuring and adjusting its rate and, of course, by setting it correctly. So the stability of a clock is a more fundamental characteristic. A stable clock is also predictable—once set to the correct time and rate it is likely to continue reading the correct time.

Pendulum clocks

Until well into the twentieth century the most accurate clocks for scientific purposes used a pendulum as the oscillator. The idea of using a swinging object as a timekeeper originated with Galileo Galilei in the 1580s. It is said that while a student in Pisa he watched lamps swinging in the cathedral and noticed that the period of swing (which

Short pendulums swing faster than long pendulums

Figure 2.2. The period of a pendulum depends only on its length. Short pendulums swing faster than long pendulums.

he supposedly checked against his pulse) seemed to depend only on the length of the supporting chain and not on how wide the lamp swung. Perhaps this tale is somewhat more credible than the later story of his dropping assorted cannonballs off the Leaning Tower of Pisa.

All students of physics learn about the pendulum. In its simplest form it consists of a massive object, known as a bob, suspended by a lightweight string or rod and able to swing freely. If the bob is pulled to one side and released, it will continue to swing back and forth for some time until it slows down and stops. Galileo's discovery was that the longer the pendulum the longer its period (see Figure 2.2), and provided the swing was not too wide, the period remained the same. He could see that the pendulum would remain a stable oscillator if the arc of the swing were kept roughly the same. Although Galileo did not build a clock driven by a pendulum, his son Vincenzio seems to have been building a pendulum clock at the time of his father's death in 1649. A model, constructed from Galileo's drawings, can be seen at the Science Museum in London.

Dutch physicist Christiaan Huygens is credited with designing the first workable pendulum clock in 1656, and with developing a simple mechanism to correct for the slight change in period with amplitude of swing. (Huygens also takes the credit for designing the first clock to use a coiled spring as an oscillator—the forerunner of the balance wheel.)

The most attractive feature of the pendulum clock is that its rate can be adjusted merely by altering the length of the rod—a short pendulum swings faster than a long pendulum. It is possible to adjust the length until the swing of a pendulum takes precisely one second. Huygens suggested in 1664 that such a pendulum could define a new "universal measure" of length that could be reproduced anywhere on Earth. This proposal was revived by the French Academy of Sciences when the metric system was being planned in the 1790s. It turns out that the length of the "seconds pendulum" is 99.4 centimetres, remarkably (but coincidentally) close to the unit of length—the metre—that was eventually adopted in the new metric system.

Pendulums are very sensitive indeed to changes in length. If you wanted a pendulum clock to be accurate to 1 second a day, you would need to maintain the length of a seconds pendulum to within 0.02 millimetres. Such small changes happen every day simply due to the expansion and contraction of the rod with changing temperature: on warm days the pendulum lengthens and so does the period; on cold days it contracts and the clock speeds up. Clock makers devised several methods of preventing this expansion, such as using combinations of different metals, but the best was an alloy of iron and nickel, called invar, designed to expand only slightly with temperature.

A more subtle influence is the strength of the Earth's gravity. On top of a mountain, where gravity is weaker, a pendulum clock will tick more slowly than an identical clock at sea level. And because the Earth is slightly flattened—the poles are closer to the centre than the equator—the length of the "seconds pendulum" varies with latitude, even at sea level. At the poles you would need a pendulum 99.6 centimetres long; at the equator 99.1 centimetres would be enough. A seconds pendulum on the Moon, where gravity is only one-sixth that of the Earth, would be only 16 centimetres long.

But pendulum clocks do have a catch. To be an accurate and stable oscillator the pendulum must swing freely, but it cannot swing freely if it is to operate the rest of the clock. Each swing advances the clock mechanism, and a little energy is taken from the pendulum. Friction and air resistance take their toll too, so the energy of the pendulum needs to be continually replenished. This is often done by a mechanism driven by slowly falling weights—such as those in a grandfather clock—but without careful design any push on the pendulum to keep it going will also alter the period of swing.

Shortt clock

The ultimate pendulum clock, indeed the ultimate mechanical clock of any kind, was invented by a British engineer, William Shortt. The first was installed in the Royal Observatory in Edinburgh in 1921. The Shortt clock had two pendulums. The first, known as the master, swung freely in an evacuated case. Its only job was to synchronise the swing of the second pendulum, called the slave, which was housed in a neighbouring cabinet. Every 30 seconds the slave sent an electrical signal to give a nudge to the master. In return, via an elaborate electromechanical linkage, the master ensured that the slave never got out of step (Figure 2.3).

Shortt clocks were standard provision in astronomical observatories of the 1920s and 1930s, and are credited with keeping time to better than 2 milliseconds a day. Many were on record as losing or gaining no more than 1 second a year—a stability of one part in 30 million. The first indications of seasonal variations in the Earth's rotation were gleaned by the use of Shortt clocks.

In 1984 Pierre Boucheron carried out a study of a Shortt clock which had survived in the basement of the US Naval Observatory since 1932. Using modern optical sensing equipment instead of the electromechanical coupling, he measured its rate against the observatory's atomic clocks for a month. He found that it was stable to 200 microseconds a day over this period, equivalent to two to three parts in a billion. What is more, the data also revealed that the clock was sensing the distortion of the Earth due to tides from the Moon and the Sun.

As we saw in the last chapter, both the Sun and the Moon raise tides in the solid body of the Earth as well as in the oceans. The effect is to raise and lower the surface of the Earth by about 30 centimetres. Since the acceleration due to gravity depends on distance from the centre of the Earth, this slight tidal movement affects the period of swing of a pendulum. In each case the cycle of the tides caused the clock to gain or lose up to 150 microseconds.

Figure 2.3. The Shortt free pendulum clock was the mainstay of astronomical timekeeping in the 1920s and 1930s. The vacuum chamber on the right contains the freely swinging "master" pendulum which transmits electrical signals to the "slave" pendulum in the cabinet on the left. This one operated at NPL from 1927 to 1959.

Quartz-crystal clocks

Wonderful though the Shortt clocks were, they were in turn eclipsed by a radically different kind of timekeeper, the quartz-crystal clock. Quartz—largely silicon dioxide—is the second most abundant mineral on Earth, and an essential component of many kinds of rocks and sands. In its pure form it occurs in clear glassy crystals.

The principle of the quartz-crystal oscillator is similar to that of the bell. A bell when struck will ring with a distinct musical note. The note depends on the shape and size of the bell and the material of which it is made. Quartz oscillators contain a ringing crystal of quartz which produces a musical note of very high frequency—normally beyond the range of human hearing. The frequency is set by the shape and size of a thin slice of quartz crystal and is very stable.

What makes quartz so suitable for a crystal oscillator is its "piezo-electric" qualities. If you squeeze a crystal of quartz it will produce an electrical voltage between its faces. On the other hand, if you apply a voltage across the crystal it will expand or contract. This means that a vibrating crystal will produce an alternating electrical signal of the same frequency as the crystal itself. That signal in turn can be fed back to tickle the crystal and keep it vibrating. A crystal oscillator, then, consists of a crystal ringing like a bell and an electrical circuit vibrating in sympathy with it.

The vibration frequency of the crystal depends on its precise shape and size. By cutting a slice of crystal in the appropriate way, it can be made to generate the desired frequency. Typical frequencies range from a few thousand to several million vibrations each second. For the oscillator to drive a clock, these high frequencies have to be divided down to provide a series of pulses that can be counted. For modern quartz clocks and watches a popular choice for the vibration frequency is 32768 hertz. This happens to be 2^{15}, so if the output from an oscillator at this frequency is halved 15 times (which can easily be done by digital electronics), a frequency of one pulse per second is obtained.

The first crystal clocks appeared in 1927, and by the late 1930s had replaced Shortt clocks as laboratory frequency standards. Because the frequency depends on the cut of the crystal, no two quartz clocks are identical. They are also sensitive to environmental conditions, especially changes in temperature. More recent oscillators contain microprocessors which automatically adjust the output frequency to compensate for changes in temperature.

Quartz clocks are everywhere. All battery operated watches, domestic clocks, alarm clocks and travel clocks are based on quartz oscillators and many of them are so good that they do not need to be reset between battery changes. Although, as we shall see, quartz clocks have been superseded by atomic clocks for the most demanding applications, they are still of vital importance in the maintenance of accurate time.

Atomic clocks

The Oxford English Dictionary traces the term "atomic clock" back to an article published in 1938. Yet the idea of using atoms to keep time is older than that. "Any radiation frequency emitted by an atom is the ticking of an atomic clock ..." said the *Standards Yearbook* of the US National Bureau of Standards in 1928.

More than a century ago Sir William Thomson (later to become Lord Kelvin) pointed out that atoms could provide a frequency standard. In a textbook published in 1879 he wrote: "The recent discoveries due to the Kinetic theory of gases and to Spectrum analysis (especially when it is applied to the light of the heavenly bodies) indicate to us *natural standard* pieces of matter such as atoms of hydrogen or sodium, ready made in infinite numbers, all absolutely alike in every physical property. The time of vibration of a sodium particle corresponding to any one of its modes of vibration is known to be absolutely independent of its position in the universe, and will probably remain the same so long as the particle itself exists."

What did Kelvin mean by the "time of vibration" of a sodium atom?

Since the early 1800s physicists had known that the spectrum of the Sun was crossed by numerous dark lines, some of which corresponded in wavelength to bright lines seen in the spectra of flames. Robert Bunsen and Gustav Kirchhoff, working at Heidelberg, showed that these lines were characteristic of the chemical elements; each element produced its own pattern of lines at sharply defined wavelengths. Sodium, for example, had a pattern dominated by a close pair of yellow lines with a

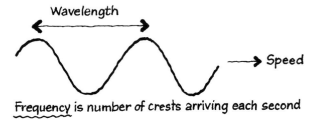

Figure 2.4. Any wave can be described by its wavelength, frequency and speed. The *wavelength* is the distance between successive crests or troughs of a wave. The unit of wavelength is the metre (m), but for light waves the nanometre (nm), which is a billionth $(1/10^9)$ of a metre, is more commonly used. The *frequency* of the wave is the number of crests or troughs that pass each second. The unit of frequency is the hertz (Hz) which is equal to one complete cycle per second. Wavelength and frequency are related to the *speed* of the wave by the simple relation:

Speed in metres per second = wavelength in metres × frequency in hertz.

The speed of electromagnetic waves is very nearly 300 000 000 metres per second.

wavelength of 589 nanometres. Dark lines appeared at that wavelength when sodium vapour absorbed light, whereas bright lines were seen when sodium vapour emitted light.

At that time no one knew how atoms emitted or absorbed light, or why the light appeared in such narrow lines of colour, but the idea that atoms somehow possessed a characteristic "vibration" that generated or absorbed a light wave quickly gained ground. By the time Kelvin wrote his textbook, light was known to be an electromagnetic wave that travelled at a high but constant speed. The yellow light waves of sodium corresponded to something in the atom that vibrated 50 trillion times a second.

As all sodium atoms were the same, Kelvin saw that their yellow light would always appear at the same fundamental frequency, so defining a period of vibration that could serve as a "natural standard" of time.

And sodium was ubiquitous; whenever chemists looked at the spectra of other substances, there was often a little sodium there as well, revealed by those bright yellow lines.

But sodium was not destined to serve as a standard of time. In 1860, during their pioneering spectroscopic investigations, Bunsen and Kirchhoff discovered a new metal. It was distinguished by a pair of blue lines and they called it caesium from the Latin for "bluish-grey". Caesium, rather than sodium, would ultimately fulfil Kelvin's predictions of an atomic time standard.

Caesium is strange stuff. It's a silvery metal that is so soft you can cut it like butter. Dropped into water it splutters and spits violently. It melts at a mere 28°C; if you could hold it in your hand (not recommended because it reacts with moisture) it would melt into a golden liquid and drip through your fingers. Of all the stable atoms it is the biggest. Although it is one of the heavier atoms—with an atomic weight of 133, it's more than twice as heavy as iron—it is actually 30 percent less dense than aluminium.

The idea that this strange metal might form the basis of an atomic clock was first publicised by US physicist Isidor Rabi in an address to the American Physical Society in 1945. Rabi had been awarded the Nobel Prize the previous year for his work in studying the magnetic properties of atomic nuclei and he took the opportunity to suggest some applications for his discoveries. Only a partial record of Rabi's lecture exists, but it is widely reported that he spoke about using his "atomic beam" method to construct an atomic clock based on the caesium atom. We shall look at how a caesium beam clock could work shortly, but to understand what makes caesium so good for timekeeping we first need to look at a much simpler atom—indeed the simplest of all atoms—hydrogen.

Hydrogen

Our picture of the internal structure of atoms has become clear only since the 1920s, with the wonderful insights provided by quantum theory. The hydrogen atom is made of a nucleus with a positive charge and an electron with a negative charge (Figure 2.5). The nucleus is a single proton, so hydrogen really is as simple as an atom can be.

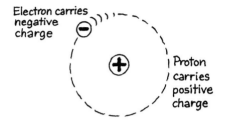

Figure 2.5. The hydrogen atom consists of a single proton and a single electron.

More than 99.9 percent of the mass of the atom resides in the nucleus, so it is reasonable to imagine the proton being stationary and the electron whirling around it like a planet moving around the Sun. The electrical attraction between these opposite charges holds the atom together, playing the role of gravity in the Solar System. The analogy should not be carried too far, since quantum theory tells us that electrons do not really behave like circling planets but more like clouds enveloping the nucleus. One consequence of this is that electrons cannot orbit wherever they like. Only certain orbits are permitted, at fixed distances from the nucleus.

The single electron of hydrogen is normally found in its closest permitted orbit—or energy level—0.053 nanometres from the proton. When the electron is in this orbit the atom is said to be in its ground state. If the electron is given a sufficient boost of energy, it can be kicked up to the next level. There it will eventually drop down again to the ground state, emitting a burst of electromagnetic radiation of precisely the same energy as it absorbed. This burst of energy is called a photon. Atoms always emit or absorb radiation in discrete photons of energy rather than continuous streams of waves. The energy carried by a photon is directly proportional to the frequency of the electromagnetic wave and inversely proportional to its wavelength. So high-energy photons have high frequency and short wavelength, while low-energy photons have low frequency and long wavelength (Figure 2.6).

Since all hydrogen atoms are identical (with a few exceptions) they will emit and absorb identical photons. If you have ever seen a photograph of a nebula—the Orion Nebula, for instance—you may have

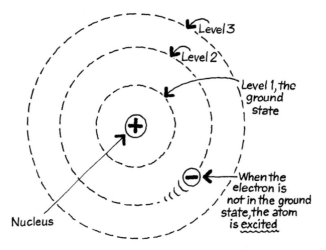

Figure 2.6. The electron can only occupy certain energy levels within the hydrogen atom. It can move to a higher level by absorbing a photon and can drop to a lower level by emitting a photon. The lowest level is known as the ground state and the higher levels are called excited states.

noticed a pink glow. That glow is caused by red photons emitted when an electron drops down from the third to the second energy level in the hydrogen atom. The red photon has a precise frequency which is the same for all hydrogen atoms. So in principle we could heat up some hydrogen atoms until they glow and then—in the spirit of Kelvin's suggestion— use the red light as a frequency standard. But remember that a clock needs both an oscillator and a counter. The hydrogen atom provides an oscillator, it is true, but what about the counter? Those red photons have a frequency of more than 10^{14} hertz, as do all the photons of visible light that are emitted from all kinds of atoms, not just hydrogen. How do you keep track of a clock that ticks 10^{14} times a second?

This was the problem that faced physicists from the 1940s onwards when they began to look for ways of making a practical atomic clock. The technology of the era offered no possibility of using visible light as a frequency standard because there was no means of counting the

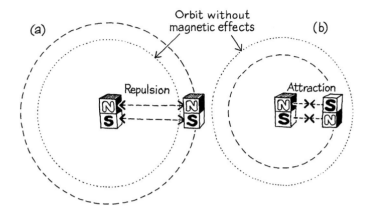

Figure 2.7. Both the nucleus and the electron of hydrogen behave like tiny bar magnets. In (a) the magnets are aligned in the same direction and the repulsive force between them pushes the electron further from the nucleus. In (b) the electron has flipped over so that the magnets point in opposite directions. They now attract each other and the electron is drawn inwards. The small difference in energy between the two orbits corresponds to a photon of wavelength 21 centimetres and frequency 1420 megahertz.

oscillations. Even today such things are at the cutting edge of research, as we will see in Chapter 8. What they needed was an atom which emitted and absorbed photons of much lower and more manageable frequencies, more like radiowaves in fact.

There is a way to get radiowaves from atoms, even from the hydrogen atom. Both the electron and the proton are magnetic. They can be thought of as having a north pole and a south pole, just like tiny magnets. In the hydrogen atom only two orientations of the magnets are stable: either they point in the same direction or in opposite directions. If the north poles are in the same direction the magnets will tend to repel each other, moving the electron slightly further from the nucleus, so loosening it. If the electron is "upside down", so to speak, then the magnet points the other way and attracts the electron somewhat more strongly, pulling it in towards the nucleus (Figure 2.7).

This means that the hydrogen's electron can be in two slightly different levels, of slightly different energy, within the ground state. These two levels are known as the "hyperfine" states. By flipping over, so that its north pole points the other way, the electron can jump from one hyperfine state into the other. The difference in energy is very small and corresponds to a radiowave photon of wavelength 21.1 centimetres, equivalent to a frequency of 1420 megahertz. The atom can change energy from the lower state to the upper state by absorbing a photon, and can return to the lower state by emitting a photon.

This so-called "spin–flip" transition is well known to radio astronomers because it is the principal means by which they can trace the presence of clouds of hydrogen atoms in the Milky Way and other galaxies. The 21-centimetre emission was predicted by Dutch astrophysicist Henk van de Hulst in 1944 and finally discovered by US and Australian astronomers in 1951.

Back to caesium

Hydrogen is the only atom with a single electron, but there are other elements which resemble it in an important way. A group of elements known as the alkali metals possess a single electron orbiting at some distance outside one or more inner shells of more tightly bound electrons. These metals are lithium, sodium, potassium, rubidium, caesium and francium (Figure 2.8).

The electrons in the inner shells are paired off, so that their magnetic fields cancel each other out. Only the nucleus and the single, outermost electron contribute to the magnetic state of the atom. The electron behaves in many ways like the single electron in hydrogen, with a pair of magnetically induced hyperfine states and the same kind of spin–flip transition as hydrogen itself.

Of all the alkali atoms, caesium is the prime candidate to make an oscillator for an atomic clock. The spin–flip transition occurs at a frequency of 9193 megahertz, equivalent to a "vibration" of almost 10 billion times a second. The frequency of this "clock transition" is the highest of all the alkali metals and can be measured more precisely than any of the others. It corresponds to a wavelength of 3.26 centimetres, which lies right in the middle of the microwave radio spectrum. These

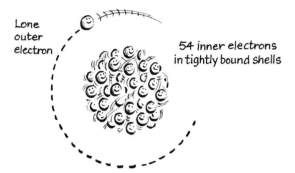

Figure 2.8. All but one of caesium's 55 electrons are tightly bound to the nucleus and usually play no part in reactions. The remaining outer electron is loosely bound, and is responsible for the properties that make caesium so useful for frequency standards.

wavelengths are used for radar, and in the late 1940s there was plenty of war-surplus radar equipment waiting to be used.

All naturally occurring caesium atoms are identical. In contrast, about one in a thousand hydrogen atoms has a neutron attached to the single proton. This form—or "isotope"—of hydrogen is known as deuterium and it has twice the mass of ordinary hydrogen. Many elements come in mixtures like this. Chlorine, for example, is made of 75 percent of atoms with 18 neutrons and 25 percent with 20 neutrons. Of the alkali metals, only sodium and caesium occur in a pure form without isotopes. Every naturally occurring caesium atom has 55 protons and 78 neutrons, giving it an atomic mass of 133. This means that every caesium atom will emit or absorb spin–flip photons at precisely the same frequency.

Caesium has still more advantages. Its low melting point means it is easy to form a vapour of caesium atoms. Once in a vapour, its high atomic mass means it moves at only half the speed of air molecules, and slow speed is highly desirable as we shall see. And apart from its bigger brother francium—which is so unstable that its atoms decay in less than half an hour—caesium is the biggest of all the atoms. As a more or less direct consequence of this, its outer solo electron is only loosely attached. It is easier to knock an electron off a caesium atom than any other atom, and this makes it easy to detect.

So following Rabi's address in 1945 it was obvious to all that the way towards an atomic clock lay in exploiting the unique properties of the caesium atom.

Ammonia clock

As fate would have it, the first atomic clock was not based on caesium at all. It was not even based on an atom, but on a molecule. Harold Lyons, a physicist working at the US National Bureau of Standards (NBS) in Washington DC, chose to design a clock based around the ammonia molecule.

The ammonia molecule is made of four atoms; one nitrogen and three hydrogens. You can imagine the nitrogen sitting at the apex of a triangular pyramid with the three hydrogens forming the base. Two forms of the ammonia molecule are possible, and they can be thought of as having the nitrogen either at the top of the pyramid or at the bottom of an inverted pyramid. Any ammonia molecule can change from one form into the other by the nitrogen atom suddenly jumping through the triangle formed by the hydrogens (Figure 2.9). For subtle reasons too complex to explain here, jumping one way requires a slight input of energy and jumping the other way results in an identical loss of energy. The energy difference is equivalent to a photon of frequency 23 870 megahertz or wavelength 1.26 centimetres. Like the caesium clock transition, this is in the microwave region of the spectrum, the sort of waves used in radar and which Lyons and his colleagues had worked with during the war. NBS had been charged with developing microwave frequency standards to support the wartime development of radar, which was then being extended to higher and higher frequencies. The ammonia "inversion" transition itself was by then well known and had been studied since the early 1930s.

Ammonia gas is usually made up of a 50:50 mixture of the two forms of the molecule, with constant interchange between them. But if a beam of microwaves of the correct frequency were to be directed through the gas, half the molecules would absorb the photons and flip into the other form.

Nitrogen

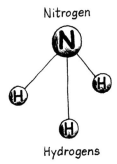

Hydrogens

Figure 2.9. The ammonia molecule is made of a nitrogen atom attached to three hydrogen atoms forming a tetrahedron. When exposed to microwaves of the correct frequency, the nitrogen will flip through the triangle formed by the hydrogens and "invert" the molecule.

At the heart of the NBS clock was a 30-foot long rectangular copper pipe (a "waveguide") filled with ammonia gas at low pressure. Microwaves were fed into one end and a receiver picked them up at the other. When the frequency of the microwaves was far from the inversion frequency the waves passed through unhindered. When the frequency was near the inversion frequency the gas absorbed the microwave photons and the receiver registered a fall in the energy emerging at the other end. The trick was to adjust the frequency of the transmitter until the signal at the other end was as low as possible. Then you knew that the transmitter was operating exactly at the inversion frequency of 23 870 megahertz.

This frequency is a fundamental property of the ammonia molecule, not the construction of the device. The rate of a pendulum clock depends on the precise length of the pendulum, the rate of a quartz-crystal clock on the precise cut of the crystal, but the rate of an atomic clock depends only on the nature of the atoms or molecules used to provide the frequency standard.

The clock itself attracted much public interest when unveiled in January 1949, not least because it looked the part (Figure 2.10). Lyons had installed the equipment in two unremarkable racks but mounted on top was the waveguide, now gold-plated, coiled around an ordinary electric clock face. It caused a press sensation.

Figure 2.10. The world's first atomic clock is unveiled at the US National Bureau of Standards in 1949. It was based on the absorption of radio waves by ammonia gas at a frequency of 23 870 megahertz. NBS director Edward Condon (left) examines a model of the ammonia molecule while the clock's designer, Harold Lyons, looks on.

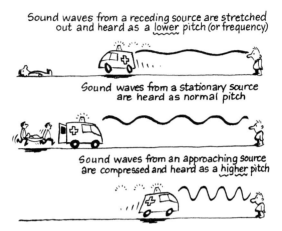

Figure 2.11. The Doppler effect. Christian Doppler, an Austrian physicist, discovered in 1842 that the frequency of sound waves heard from a moving object depends on the relative speeds of the source and the listener. The siren of an ambulance, for example, appears to have a higher pitch (frequency) when it is coming towards us than when it is going away. This is simply because the motion of the ambulance compresses the waves as it approaches and draws them out as it recedes. Doppler showed that the fractional change in frequency (and hence wavelength) is directly proportional to the relative speed of the source and the listener. The greater the speed, the more the frequency will appear to change. The Doppler effect applies to all kinds of waves, including electromagnetic radiation. Light waves, in particular, are said to be "red-shifted" when they are lengthened and "blue-shifted" when they are shortened.

Unfortunately the clock did not live up to expectations. Though an improved version eventually attained a stability of two parts in 10^8, this did not offer much of an advantage over existing quartz-crystal clocks and was no more stable than the rotation of the Earth. When first announced it could only run for a few hours at a time, though this was eventually extended to a few days.

The main problem was that the absorption frequency was rather broad. Instead of absorbing microwaves at the expected sharp, well-defined inversion frequency, the molecules absorbed over a broad range of frequencies. The reason for this is the Doppler effect (Figure 2.11). At normal room temperature gas molecules dash about in all directions at hundreds of metres per second. If we could look down the waveguide of the ammonia clock we would see about half the ammonia molecules going away from us and about half coming towards us. If we then sent the microwave beam down the waveguide those molecules moving towards the beam would see its waves blue-shifted, and its frequency would appear slightly higher. Likewise those molecules moving away would see a red-shifted beam of lower frequency. That means that, even if the radio beam is tuned precisely to the inversion frequency, many of the molecules will not absorb a photon because they see the frequency as either too high or too low. On the other hand, if the radio frequency is slightly different from the inversion frequency some molecules will still absorb because their speed has shifted the frequency to the correct value. The effect of this is that radiation is absorbed in a band of frequencies around the inversion frequency, making it less precise. There was nothing wrong in principle with using ammonia to make an atomic clock, but the wandering molecules blunted the precision of the device.

Caesium beams

There is no doubt that the NBS ammonia clock was the world's first atomic clock, even though it proved to be a cul-de-sac in terms of the technology. The road towards an atomic clock based on caesium finally proved much more rewarding, as Rabi had foreseen. It is not so clear who should take the credit for the first caesium atomic clock, both because there were three machines in contention for the title and because the term "atomic clock" was open to interpretation.

In 1948 Lyons had appointed Polykarp Kusch, Rabi's colleague at Columbia University, to look into the possibility of building a caesium clock at NBS. Kusch (also destined for a Nobel Prize) outlined the design the following year, and in 1951 the NBS experimental caesium beam standard was operated for the first time. But the apparatus was temperamental and after 1952 little progress was made in the face of budgetary cuts. In 1954 the machine was dismantled as Lyons's group was relocated from Washington DC to the new home of the NBS Time and Frequency Division in Boulder, Colorado.

With the NBS programme stalled, a second strand of the story was in progress at the Massachusetts Institute of Technology (MIT). Jerrold Zacharias, another physicist who had distinguished himself in work on wartime radar, was intent on designing a caesium beam clock that was not only transportable but could be manufactured and sold commercially. Since no one had ever constructed such a clock, even an experimental model, this was a high ambition. Yet by the summer of 1954 a prototype of a caesium beam standard was working at MIT but it would be another two years before a working clock became available.

The decisive advance came from unexpected quarters, a small and relatively inexperienced team led by Louis Essen at the UK National Physical Laboratory. Essen had come to NPL as a newly graduated physicist in 1929, joining the Electricity Department to work on frequency standards. By 1938 he had perfected an improved kind of quartz-crystal oscillator in which the crystal was cut in the form of a ring, and which soon became known as the "Essen ring" frequency standard. During the war Essen worked on microwave technology and used what he had learned to devise a new method of measuring the speed of light. His final value, announced in 1950, came out to be 299 792.5 kilometres per second, some 16 kilometres per second faster than was then accepted, but in good agreement with the ultimately agreed value of 299 792.458 kilometres per second.

With this background in precision timekeeping, microwave technology and meticulous measurement, Essen had been following with interest the atomic clock developments at NBS and MIT. He had seen the ammonia clock (writing a popular feature about it for *Practical Mechanics* magazine) and had visited the main players in the US effort to develop atomic frequency standards. Like them, he knew that the way forward lay with caesium beams of the type discussed by Rabi. Early in 1953 he returned from a visit to the US determined to build a caesium beam frequency standard at NPL. With a modest budget, Essen set to work with his colleague, John ("Jack") Parry. Even though neither of them had experience with atomic beams, they were skilled experimentalists and by June 1955 the NPL caesium standard was up and running.

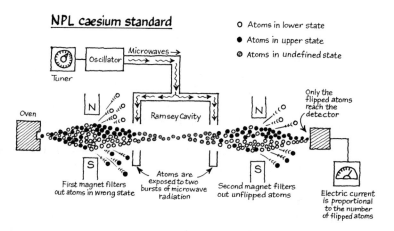

Figure 2.12. Principle of the NPL caesium standard. Caesium atoms from the oven emerge in a stream at around 200 metres per second. They are a mixture of equal numbers in the upper hyperfine state (black dots) and lower hyperfine state (white dots). At the first magnet the atoms in the correct state are focused into a beam while the others are deflected away. The beam passes through the arms of a Ramsey cavity in which they are twice exposed to microwaves near the transition frequency. On emerging from the cavity the beam contains atoms which have been "flipped" into the opposite state and those which have not. The unflipped atoms are deflected away by the second magnet, while the flipped atoms are focused onto the detector. The greater the proportion of atoms arriving at the detector, the closer the microwave frequency to 9192 631 770 hertz. The NPL machine measured 1.2 metres from oven to detector, the gap between the poles of the magnets was only 2 millimetres and the deflection angles were less than half a degree.

Figure 2.12 shows the principle of the NPL caesium beam frequency standard. It was based on the proposals of Kusch, which in turn owed their origins to Rabi's vision of an atomic clock. Since all subsequent caesium beam machines have worked on much the same principles—until the 1990s at least—we'll look into its operation in some detail.

The object is to tune a very stable quartz-crystal oscillator to the hyperfine frequency of caesium and to keep it there. The working parts of the machine are contained within an evacuated tube since all traces of air have to be removed if the caesium beam is to fly straight. At the left is a small, electrically heated oven containing less than a gram of caesium heated to about 200°C. The hot atoms of caesium stream from a narrow slit in the side of the oven at a speed of around 200 metres per second. This is the "beam" of caesium atoms.

These atoms are a mixture; about half will be in the lower hyperfine state and half in the upper state, and the first job is to separate them. This is done by passing the beam through a specially shaped magnet, a technique devised by Otto Stern and Walther Gerlach in the early 1920s to study the structure of atoms. Since the caesium atoms themselves are magnetic, they are deflected by the magnetic field according to their energy states and emerge from the magnet in two slightly different directions, one for the lower state and one for the upper state. The geometry ensures that atoms in the upper part of the beam are in one state and those in the lower part in the opposite state.

The atoms are then ready to be flipped. Caesium atoms are like very precisely tuned radio receivers. They will ignore passing waves of the wrong frequency but respond strongly to waves of the right frequency, namely 9193 megahertz. An atom in the lower state hit by a photon will absorb it and flip to the upper state. An atom in the upper state hit by a photon will release an identical photon and flip to the lower state. In each case the outer electron is turned over by the incoming wave and changes the state of the atom.

In Rabi's original method the atoms pass through a long tube in which they are exposed to microwaves. The longer they remain in the tube the more accurately the transition frequency can be defined. But the NPL machine, following the ideas being implemented at NBS, used a new method which had been devised by Norman Ramsey of Harvard University in 1949, and ultimately led to his being awarded a Nobel Prize 40 years later.

Ramsey realised that there was no need to pass the atoms through a long tube full of microwaves. He saw that the transition could be achieved just as effectively if the atoms were subjected to microwaves in two short bursts rather than a single long exposure. Not only does this method simplify the engineering—it was very difficult to control the exposure in a long tube—but it actually increases the sensitivity of the machine. In the NPL machine, and all others since, the microwaves are fed into a waveguide—now known as a Ramsey cavity—which divides into two arms to form a U-shape. The caesium beam passes first through a hole in the end of one arm of the U and then through a similar hole in the other arm. At each pass through the cavity the atoms receive an identical burst of microwaves. If the frequency of the radiation corresponds to the transition frequency, the first burst puts the atoms into a ghostly quantum mixture of the two states and the second completes the transition to the opposite state. The sharpness of the transition frequency is now proportional to the length of time the atoms are coasting in limbo between the two bursts of radiation. The longer the coast, the more accurately the frequency is defined.

Because the atoms are all streaming in nearly parallel lines and because the radiation is directed at right angles to the beam, the microwaves hit the atoms from the side, so to speak, rather than head on. That means there is no Doppler shift in the wavelength of the radiation seen by the atoms and so, unlike the ammonia clock, the frequency remains sharp.

Emerging from the second arm of the Ramsey cavity the beam now consists once again of two kinds of caesium atom: those that have been flipped and those that have not been flipped. A second magnetic filter separates them in the same way as the first. The geometry of the beams ensures that the flipped atoms are focused onto a detector while the now unwanted, unflipped atoms are sent to oblivion.

The detector is simply a red-hot wire in front of a metal plate. Remember that the outer electron of caesium, the one that confers its magnetic properties, is weakly bound to the rest of the atom. The temperature of the wire is sufficient to knock off that electron while letting the atom itself rebound as a positively charged ion to be collected by the negatively charged plate. This causes a current to flow through the plate and the strength of the current is proportional to the number of flipped atoms coming through the machine.

As with the ammonia clock, the trick is to tune the microwave transmitter feeding the cavity until the current from the detector reaches a peak. At this point you know that the maximum possible number of atoms are being flipped and that can only happen when the microwave frequency is the same as the caesium transition frequency. (When the NPL machine was first operated, the unflipped atoms were detected, which meant that the current was least when the frequency was at resonance. This arrangement was later changed.) In later atomic clocks the microwave frequency was automatically locked to the caesium frequency, but in the NPL machine the adjustment had to be done manually.

Once the machine was running (see Figure 2.13), one of Essen and Parry's first tasks was to measure the length of the second in terms of the caesium transition. They found that 1 second (a mean solar second, defined by the Royal Greenwich Observatory and broadcast as radio time signals) was equal to 9 192 631 830 cycles of the caesium frequency, with an uncertainty of ten cycles either way. No other physical quantity had ever been measured so accurately.

Even if the NPL machine was not the first caesium beam standard, it is widely recognised as being the first that was operated successfully and the first that measured a second. Was it a clock? By our earlier definition of a clock as having an oscillator and a counter it was not. The first published account of the machine, in the scientific journal *Nature* for 13 August 1955, described it as "an atomic standard of frequency and time interval"—an exceptionally stable oscillator, certainly, but it did not count seconds and nowhere in the paper did Essen and Parry describe it as a clock.

Although the machine was never operated as a clock in the normal sense of the word—it did not tell the time—it was used at intervals to calibrate the already very accurate quartz-crystal clocks at NPL and thereby to establish an atomic time scale of great stability. When first operated the standard was accurate to one part in a billion, but this was soon improved to one part in 10 billion. With periodic calibrations from the caesium standard the quartz-crystal clocks were then accurate to two parts in 10 billion. In effect, the quartz clocks acted as the face of the atomic clock.

In a very short time the NPL machine did become known as the world's first caesium atomic clock and Essen was soon using the term himself. But whether we choose to call it a clock or not, it would soon

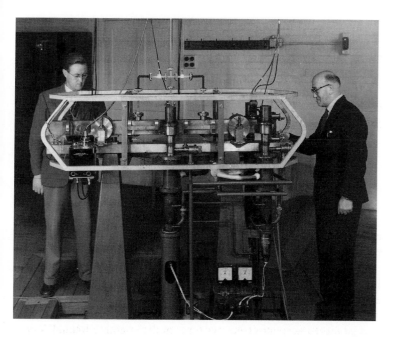

Figure 2.13. Louis Essen (right) with Jack Parry and the world's first operational caesium clock. The long box in the centre is the beam chamber which contains the working parts. Caesium atoms emerge from an oven at the right and pass to a detector at the left. The circular objects near each end are the coils of the deflecting magnets. Microwaves enter through the vertical waveguide at the top of the picture and pass into the U-shaped Ramsey cavity which guides them into the beam. The white framework is made up of coils of wire which shield the equipment against magnetic fields. The three vertical cylinders inside the framework are vacuum pumps.

make a decisive contribution to timekeeping, as we shall see in the next chapter. As Essen later wrote, "We invited the Director to come and witness the death of the astronomical second and the birth of atomic time. And it was indeed the birth because much to our surprise it was another year before any clocks were working in the USA."

3

ATOMIC TIME

"The times they are a-changin'..." Bob Dylan, 1964

A physical second?

Accompanying the announcement in *Nature* of the success of the NPL caesium standard was a note by Sir Edward Bullard, the Director of NPL. He pointed out that observations of the Moon over a period of four years would be necessary to determine Ephemeris Time to the accuracy achieved by the NPL atomic clock in a matter of minutes. "The natural way of escape from this difficulty," he wrote, "is to define a 'physical second' in terms of the natural period of the caesium atom, choosing the numerical value so that it agrees as well as may with the current estimates of the second of Ephemeris Time." Bullard chose his words carefully, for he intended "physical second" to be understood as being in distinction to "astronomical second". He was articulating what many physicists had been thinking, and no doubt some astronomers too, that the time was near when timekeeping would be taken out of the hands of astronomers and handed over to physicists.

Within three months of operating the world's first caesium clock, Louis Essen was in Dublin to attend the General Assembly of the International Astronomical Union. As we have seen, the IAU was in the process of engineering a seamless transition from Universal Time, and all its irregularities, to the smoother running Ephemeris Time. One of the items on the agenda was a motion that, if passed, would finally confirm the adoption of ET as the basis for world timekeeping. Essen believed that this process had already been rendered futile by the advent of the atomic clock, and that timekeeping in future would be based on atoms, not on the motions of the planets.

When the second was defined in terms of the solar day, astronomers could make observations of the transit of standard stars and determine UT to a few milliseconds literally overnight, with an accuracy of a few parts in 10^8. With ET as the official time scale it would only be possible to assign a precise time to an event after laborious observations and computations stretching over several years. Both in its accuracy and its immediate availability, the newly born atomic time was far superior to Ephemeris Time, yet the astronomical community was about to commit itself, and through CPGM the rest of the world, to a timekeeping system that could be realised only with difficulty after long periods of observation.

After reporting on the successful operation of the NPL atomic clock, Essen suggested that it would be wise to defer a final decision on introducing ET until agreement could be obtained on the definition of an atomic unit of time which would certainly be required in the not too distant future. But it was not to be. The IAU voted to proceed with Ephemeris Time. ET was here to stay, for the time being anyway, and Essen knew that the first step towards its eventual replacement by atomic time would be to measure the length of the new ET second in terms of the frequency of the caesium transition. The NPL team had already calibrated their atomic clock against the mean solar second, with the help of time signals from the Royal Greenwich Observatory, but the ET second was altogether a more formidable challenge.

From UT to ET

As well as settling on Ephemeris Time, the Dublin IAU meeting decided, in an attempt to come to terms with the fluctuations in the Earth's rotation, that there were now to be three forms of UT. The kind of UT that could be measured directly by observing star transits, and derived directly from sidereal time, was called UT0. Because of polar wobble, which altered the latitude and longitude of the observatory, UT0 differed slightly from place to place. When UT0 was corrected for polar wobble it became UT1, which was the same everywhere on Earth. UT1 was now seen primarily as an angle, a measure of the Earth's orientation in space rather than a time—this was the version of UT to be used by navigators. Finally, when UT1 was corrected for the seasonal variation in rotation speed, the result was UT2. UT2 would be the modern equivalent of GMT, an approximation to a smoothly flowing time scale based on the rotation of the Earth. UT2 was to be the basis of civil timekeeping, even though the length of the UT2 second would still be changing due to tidal friction and the non-periodic variations in the Earth's rotation.

In Dublin, Essen met William Markowitz, of the US Naval Observatory, who had been following the NPL work with interest. As Director of the Time Service Division at the world's leading centre for astronomical timekeeping, Markowitz had already turned his attention to how ET was related to the new UT2 and how it could be determined in practice.

Each of the planets, remember, is like one hand of a multiple-handed clock, sweeping out time at its own rate but geared together to read ET. If we keep proper count of its movements, any hand will tell the time, but just as the second hand of a clock tells the time most precisely, so the faster moving planets are the most accurate indicators of ET. The Moon completes its orbit faster than any of the other major bodies in the Solar System and so, like Edmond Halley and his contemporaries in the eighteenth century, it was to the Moon that astronomers now turned to tell the time. That required precise measurements of the position of the Moon relative to the background of stars. Knowing the position of the stars one can then calculate the precise position of the Moon in its orbit, and from there, work out the ET at which the Moon was at that position. If one were public spirited, as of course were the staff of USNO, one could then publish a table of corrections from UT2 to ET as a service to the astronomical community.

Markowitz was well aware of the practical difficulties in using ET and, like Essen, he saw the urgency of relating the duration of the ET second to the second of atomic time. Markowitz had the means to measure Ephemeris Time, but he did not have an atomic clock. Essen had an atomic clock, but he did not have the means to measure Ephemeris Time. Thus a collaboration was born.

Although the IAU meeting had given a lukewarm reception to Essen's atomic clock and saw no reason to delay adopting ET, Markowitz persuaded the assembled astronomers that, far from being a threat, the atomic clock was now just what they needed to make ET available quickly and accurately. But before that could happen, it would be necessary to establish the relationship between the ET second and the vibrations of the caesium atom. When that was done, atomic clocks would read ET to far greater accuracy than any quartz clock reading UT2. Having won an IAU resolution supporting their project, Markowitz and Essen began work.

The task that the NPL–USNO team set themselves was twofold. First they would count the number of caesium cycles in the UT2 second over a period of three years or so. At the same time they would make observations of the Moon to correct the UT2 measurements to ET and so determine the length of the ET second in terms of the caesium frequency.

USNO had long experience of determining UT—it was their job after all—and they had set up specialised telescopes for that purpose, known as photographic zenith tubes, at observing stations in Washington, DC, and Richmond, Florida. It was a routine matter for USNO to establish a UT2 time scale and maintain it with the help of their quartz-crystal clocks. But how could the second of UT2 as realised at USNO be made available to NPL across the Atlantic?

In Chapter 5 we will have more to say about how time can be transferred from one place to another, but an important method is to employ radio stations that broadcast time signals. In the 1950s a short-wave station operated by NBS and called WWV (then broadcasting from Greenbelt, near Washington DC) transmitted pulses at nominal 1-second intervals. Once a month NPL measured the interval between the pulses in terms of the caesium period using their crystal clocks calibrated by the caesium standard and USNO measured the interval in terms of UT2 using their crystal clocks calibrated by the photographic zenith tubes.

The precise intervals of the broadcast signals did not matter: what mattered was that they were being measured simultaneously in terms of UT2 and the number of caesium cycles. By exchanging their measurements of the time signals, the NPL–USNO collaborators could then calculate the length of the UT2 second in terms of the caesium frequency.

As expected, the second of UT2 measured by the caesium standard was by no means constant (Figure 3.1). Even with the seasonal fluctuations smoothed out by averaging over a year, the length of the second

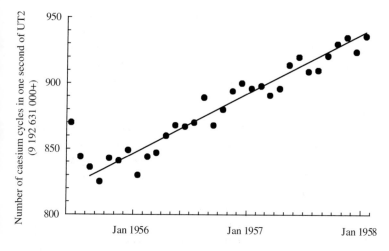

Figure 3.1. Measurements made at NPL and USNO in the late 1950s (black dots) appeared to show that the frequency of the caesium transition was increasing steadily when measured against UT2—the dots fall close to a straight line. In reality, the rotation of the Earth was slowing down and the second was becoming longer. More caesium cycles were being completed each second, giving the illusion of increasing frequency. Compare this diagram with Figure 5.3.

increased steadily from the summer of 1955 until the project ended three years later. But this lengthening of the day was 50 times greater than could be accounted for by tidal friction; a tantalising glimpse of the rich research opportunities waiting to be exploited by those in possession of atomic time.

The Moon camera

While the second of UT2 was being related to atomic time at NPL, Markowitz's team was relating UT2 to Ephemeris Time by observations of the Moon. The traditional means of measuring the precise position of the Moon was (and is) by timing the disappearance ("occultation") of a star as the Moon passes in front of it. Stars are so distant that they appear as pin-points of light so the disappearance is almost instantaneous. If the time of the occultation can be measured accurately, the position of the edge of the Moon at that instant is determined very precisely. But occultations of bright stars are infrequent and a single measurement was not sufficient to locate the centre of the Moon. Ideally, one would want to take photographs of the Moon and then measure its position with respect to several surrounding stars.

But there are problems in taking photographs of the Moon from which such measurements can be made. In the first place, the Moon is extremely bright compared with the background stars. To photograph the Moon you need a short exposure, but then the stars would not show up. To photograph the stars you need a much longer exposure, but then the extreme brightness of the Moon would wash out its image and render accurate measurement impossible.

The second and more serious problem concerns the very reason why the Moon was chosen to measure ET, namely its rapid motion among the stars. Astronomical telescopes are designed to compensate for the rotation of the Earth by being driven in the opposite direction to track stars as they move across the sky. The Earth makes one revolution eastwards, 360 degrees, in 24 hours, so the telescope needs to move westwards at a rate of 1 degree every 4 minutes to keep on target. By this means astronomers can take long exposure photographs of faint objects, often for several hours at a time. The Moon, like all astronomical objects, shares this general westerly motion but its orbital speed carries it eastwards among the stars at a rate of 13 degrees a day or about half a second of arc every second of time. Astronomers can take pictures of the Moon by adjusting the tracking speed to match the motion of the Moon rather than the stars. But it is not possible to track both at the same time. If the telescope is set to track the stars, then the edge of the Moon blurs very quickly. If the telescope is set to track the Moon, the stars spread out into trails. The reason the Moon is such a good marker of Ephemeris Time—its rapid motion—is the same reason it is so hard to photograph.

Since photography was first used to record the Moon's position around the turn of the century, camera designers had come up with several solutions to these twin problems of the Moon's brightness and motion. This usually involved devices to restrict the light from the Moon

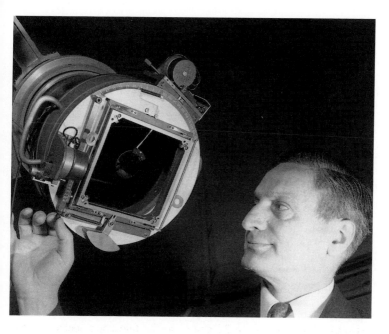

Figure 3.2. William Markowitz, of the US Naval Observatory, with the Moon camera he designed for the measurement of Ephemeris Time. In this picture the back of the camera has been removed to reveal the circular filter which slowly tilts to prevent the image of the Moon drifting against the stars.

while giving the stars a full exposure, either by placing a disk in front of the telescope to block out the Moon after a short exposure, or a special shutter near the photographic plate itself. For high precision work these techniques were not ideal, because the Moon and the stars were not photographed at the same time. Any slight error in the telescope tracking during that period could result in the image of Moon being slightly displaced from those of the stars.

Markowitz's solution was known as the dual-rate Moon camera (Figure 3.2). He designed it in 1951 with the express purpose of determining ET and it came into use at the US Naval Observatory in June of the following year.

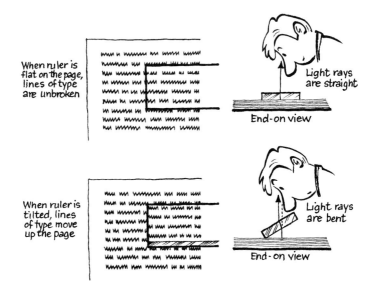

When ruler is flat on the page, lines of type are unbroken

Light rays are straight

End-on view

When ruler is tilted, lines of type move up the page

Light rays are bent

End-on view

Figure 3.3. A plastic ruler can be used to demonstrate the principle of the Moon camera (see text).

To see how the Moon camera worked, take an ordinary transparent plastic ruler (flat rather than bevelled) and lay it on the page of this book, parallel to the lines of type, so that one end is about halfway across the page (Figure 3.3). If you look vertically down on the ruler, you can see how the lines continue unbroken beneath it. Now tilt the ruler about its length, while still looking vertically downwards. The lines are no longer continuous. If you lift the lower edge of the ruler, the lines of type move up the page. If you lift the upper edge they move down. The effect is only slight, and you have to tilt the ruler by quite a large angle before there is any noticeable shift. The reason for this is that the light from the page is bent on passing through the tilted ruler. When the ruler is flat the light passes straight through, but when the ruler is tilted, refraction ensures that the light takes a staggered route through the ruler and emerges in a slightly different position.

How does this help us photograph the Moon? Inside Markowitz's camera was a photographic plate 16 centimetres square. In front of it was mounted a circular filter—a disk of dark glass—1.8 millimetres thick and just big enough to cover the image of the Moon on the plate. As the exposure progressed the Moon moved away from its initial position, but as it did so the filter started to tilt. Just as the tilting ruler moves the lines of type in our experiment, the tilting filter moves the image of the Moon. The rate and direction of tilt was calculated to move the image of the Moon at the same speed but in the opposite direction to its motion on the sky. The result was that the image of the Moon remained precisely in the same place on the plate for exposures up to 20 seconds, the effective instant of the exposure given by the moment at which the filter was parallel to the plate. The darkness of the filter—it passed only 0.2 percent of the Moon's light—ensured that the Moon's image was not overexposed.

The engineering tolerances of the Moon camera were severe, in Markowitz's own words. For precision work, he knew that the tracking had to keep within 0.1 seconds of arc of the correct position. At that time the drive on the telescope he was planning to use was not stable enough, so he decided to keep the telescope stationary and build the drive into the camera itself. In operation he would line up the telescope on the Moon, switch off the drive, and let the camera's internal motor slowly move the plate holder inside the camera to follow the stars. A 10-second exposure was enough to record several measurable stars.

Markowitz also designed a special "engine" to measure the glass plates. The plate was mounted on a carriage that could be moved in two directions, by calibrated screws, and also rotated. Using a microscope with cross hairs, a technician measured the position of the edge of the Moon at 30 points and these were used to calculate the precise position of the centre. The positions of the known stars on the plate were also measured. From the coordinates of the stars in a catalogue the precise position of the Moon on the sky could then be obtained. Finally, in accordance with standard astronomical practice, the position was adjusted to the position of the Moon as seen from the centre of the Earth.

Now Markowitz and his colleagues had a series of precise positions of the Moon, each for a known UT2. The final step was to find out the ET for each photograph. To do that, they consulted the *Improved Lunar Ephemeris*, calculated by the US Nautical Almanac Office. This contained predicted positions of the Moon at regular intervals of ET. As ET was now effectively defined by the motion of the Moon this book was the key to Ephemeris Time, and with its help they arrived at a precise ET for each of the Moon photographs, and so a value of the difference between UT2 and ET. They already knew the number of caesium cycles in a UT2 second—it was steadily increasing as shown in Figure 3.1—so it was now a simple matter to compute the length of the ET second in terms of UT2 and find out the number of caesium cycles in an ET second.

In 1958 the USNO–NPL team announced the results of their three-year programme. Unlike the second of UT2, the second of ET was indeed constant and was found to be equal to the duration of 9192 631 770 cycles of the caesium frequency, with an uncertainty of 20 cycles either way. This was a precision of two parts in 100 million, and it was limited not by the atomic clock but by the difficulty of measuring Ephemeris Time.

Atomic time

Markowitz had joined forces with Essen in 1955 because there were no atomic clocks operating in the US. For a year or so Louis Essen and Jack Parry had the pleasure of running the world's one and only atomic clock, but events were moving quickly. While NBS struggled to get their relocated caesium standard working again, Jerrold Zacharias at MIT was about to put his transportable caesium clock into commercial production with the National Company under the name "Atomichron". Other frequency standards were being constructed in Canada and Switzerland as well as in the US, and soon several other laboratories would join the small club of atomic timekeepers.

Some of these frequency standards were used to construct atomic time scales. Just as astronomical time scales are made by counting days, atomic time scales could be made by counting periods of the caesium frequency. When the counter reached 9192 631 770, exactly 1 second had passed. The seconds themselves were then counted up to form minutes, hours and days.

The first atomic time scale started in June 1955 and was formed by the Royal Greenwich Observatory using data from the NPL caesium standard. The Bureau International de l'Heure (BIH), operating under the auspices of the IAU to maintain a world system of astronomical time, quickly followed suit with a scale called AM. Based initially on data from NPL it later took in caesium standards in the US, Switzerland and France, and was disseminated by publishing corrections to be applied to existing radio time signals. The widely separated atomic clocks were co-ordinated by means of long-distance, low-frequency radio transmission from navigation beacons, in much the same way as WWV was used to transfer UT2 across the Atlantic. The first US atomic time scale, called A1, was established by the US Naval Observatory in 1959. Along with the BIH scale it was defined to coincide with UT2 at 00:00 UT2 on 1 January 1958.

Of all the physical units, the second is unique in that it can be made available instantly and with considerable accuracy by radio transmissions. But when we talk about the BIH "time scale" we need to be clear that BIH was not transmitting time signals of its own: rather, it published corrections to standard time signals that would allow them to be adjusted to atomic time after the event. This meant that atomic time was only available to its full accuracy in arrears.

One of the problems plaguing ET was now inherited by atomic time. ET was set close to UT at the beginning of 1900, with the length of the ET second determined from observations of the average length of the mean solar day in the eighteenth and nineteenth centuries. Now, in the 1950s, the day had lengthened, UT2 was running about 30 seconds behind ET and the ET second was far from being 1/86 400 of the mean solar day. The ET—and soon-to-be SI—second was simply too short to match the Earth's rotation in 1958. That meant that atomic time was not only diverging from Earth time, it was racing ahead.

This posed a dilemma for the world's timekeepers. Atomic clocks, coordinated by BIH, were now producing an unending stream of ET seconds of fabulous accuracy, each identical to all the others as far as it was possible to tell. But the hard-won internationally agreed second was too short. A year contained 31 536 000 mean solar seconds, but more

like 31 536 001 ET seconds. For all sorts of good reasons, not least
to avoid confusing navigators, it was judged that UT, as disseminated
through time signals, should still be tied to the rotation of the Earth and
maintained close to UT2—the year would continue to be 31 536 000
seconds in length.

From 1 January 1960, a handful of observatories and laboratories—
the US Naval Observatory, the Royal Greenwich Observatory, the National Bureau of Standards, the US Naval Research Laboratory and the
UK National Physical Laboratory—agreed to support a new version of
UT called Coordinated Universal Time (UTC). They settled on a formula, to be administered by BIH, to maintain UTC approximately in
step with UT2. That was done by two devices. First, the length of the
UTC second would be adjusted each year so that the mean solar day
continued to have 86 400 seconds, or as nearly as could be predicted.
This meant that the pulses transmitted by radio time signals were not,
in fact, ET seconds, but something close to mean solar seconds. BIH
would monitor the rotation of the Earth—through observatories such as
USNO—and if the predictions turned out to be off the mark, BIH would
introduce corrections of a fraction of a second to bring UTC back into
step with the Earth. In other words, UTC would now and again include
a "second" that was slightly shorter or longer than usual.

At the beginning of January 1961, for example, the UTC second
was stretched by 15 parts in a billion relative to the ET second. The
first second of August 1961 was then trimmed to 0.95 seconds to prevent
UTC running too far behind the spinning Earth. In the following January
the stretching was relaxed to 13 parts in a billion, and in November
1963 UTC was paused for 0.1 seconds to allow the Earth to catch up.
Two months later the second was again extended, now by 15 parts per
billion, and in April 1964 another 0.1 second pause was introduced. By
September 1965 UTC had been paused a further five times and in January
1966 the second was stretched even further to 30 parts in a billion. At
that rate UTC ran smoothly until February 1968, when it was again held
up for another tenth of a second.

This was not an ideal way of running the world's timekeeping system. Every time UTC was adjusted in rate or paused, all the clocks in
the world were expected to do the same. In the case of radio time signals

that sometimes meant new equipment had to be installed. Calculating the interval of time between two dates became a nightmare because the varying length of the second and the irregular hiccups meant one could not simply subtract one time from another. And many questioned the logic of adjusting the length of a defined unit—the ET and SI second—to make it match an interval of time—the UT2 second—which was known to be naturally variable.

In reality, many time services continued to make their own adjustments, regardless of the announcements coming from BIH. For example, the US radio station WWVB—a sister service of WWV—broadcast what was called "Stepped Atomic Time", in which the intervals were precise ET seconds with a frequent "step" correction to keep within 0.1 seconds of UT2. And it was not at all clear who should be in charge of timekeeping. Was it the national observatories, who could see their control of time slipping away, or was it the standards laboratories who were in possession of the atomic clocks? In the UK, for example, a delicate compromise was reached: NPL was responsible for controlling the transmission frequency of broadcasts while RGO looked after the time signals.

Towards the atomic second

By the time the ET second had become the official SI unit of time in 1960, several standards laboratories were generating atomic time scales of far greater precision and accuracy than could be determined from astronomical observations. Now that the NPL–USNO project had tied the ET second to the caesium transition, an atomic clock could in a few minutes realise the SI second to an accuracy that would take over a year to be achieved from measurements of the Moon. Whether the astronomers liked it or not, the *de facto* source of ET seconds rapidly became atomic clocks, rather than the laboriously constructed time scales from astronomical observatories.

In December 1963 a CGPM committee of experts, among them William Markowitz and Louis Essen, met in Paris to discuss whether the SI second could be redefined in terms of an atomic time standard. This was the third time they had met to debate the issue. At previous meetings the committee had been in no hurry to abandon the ET definition and preferred to wait to see how atomic timekeeping would develop. Other types of atomic clock were coming on the scene, notably the hydrogen maser invented by the same Norman Ramsey whose microwave cavity had made caesium clocks practicable, and it was not yet clear which atom would ultimately provide the standard. But more than eight years after the first caesium clock, they now felt that it was time to make a move. As a result, the CGPM the following year adopted a provisional—if inelegant—definition of the SI second based on atomic time. "The standard to be employed is the transition between the two hyperfine levels $F = 4$, $m_F = 0$ and $F = 3$, $m_F = 0$ of the ground state $^2S_{1/2}$ of the atom caesium-133 undisturbed by external fields and the value 9192 631 770 hertz is assigned."

By the thirteenth General Assembly of the CGPM in October 1967 any remaining doubts had been dispelled. Atomic clocks had proved their worth to everyone's satisfaction. By an overwhelming majority the conference voted to adopt a resolution in French which is conventionally rendered into English as: "The second is the duration of 9192 631 770 periods of the radiation corresponding to the transition between the two hyperfine levels of the ground state of the caesium-133 atom." No longer was the second related to the length of the day, nor to the movements of the planets; no longer would the world look to the sky for its source of time.

So ended the era of astronomical time.

Postscript

In 1962, its job done, the NPL caesium standard was donated to the Science Museum in London. It now sits in the Time Gallery on the first floor, its skeletal frame rather out of place amid the polished wooden cases of more conventional clocks and its stillness attracting rather less attention from the bustling crowds. Its beam tube has been dissected out and displayed for our perusal like some specimen from the neighbouring Natural History Museum. You can see the little oven from where the caesium atoms streamed and the simple detector that caught them at the other end. You can see the deflecting magnets which segregated the atoms, the U-shaped Ramsey cavity which flipped them over, and the surrounding coils which protected the machine from the magnetic fields of the outside world. The atomic clock is brown now, the brass is tarnished and the varnish is discoloured. Yet this ungainly clock without a face ushered in the new era of atomic time.

4

WORLD TIME

"A man with one clock is sure of the time; a man with two clocks never is." *Anon*

The mythical world clock

If you were asked to design a world timekeeping system, how would you set about it? Perhaps your first thought would be to construct the world's best atomic clock—no expense spared—and set it up at BIPM, the keeper of the world's standards of measurement. It would tick away for eternity, distributing SI seconds to the whole world. Anyone who wanted to set their own clock to world time would simply tune in to the pulses from Paris.

The notion of a world clock has some appeal, but it also has several drawbacks. For one thing, clocks have a tendency to stop. If time is based on the rotation of the Earth this is not a problem, since the Earth itself can keep time while the clock is restarted. At one rotation each day, you would have to be very careless indeed to lose count of the Earth's "ticks". But atomic clocks are different. With more than 9 billion ticks a second, the stopping of the world clock would be a disaster from which it would be hard to recover.

And there is no doubt it would stop. Aside from unexpected breakdowns and interruptions for maintenance, caesium beam clocks eventually run out of caesium. It is boiled away at one end of the beam tube and transported to the other end. After running for a few years a caesium beam clock will need to be replenished. Even minor hiccups could disrupt world time.

Another drawback of a world clock is that atomic clock technology is moving very fast. The best frequency standards are state-of-the-art devices. Their designers are constantly improving them, trying out new ideas and new techniques. If we built a world clock now, it would be out of date this time next year. How would we keep it the world's best clock without stopping it now and again to upgrade it?

But there is an even more fundamental objection to having a world clock. How would we know it was reading the right time? It may be the best clock in the world, but suppose it developed some minor fault that caused it to read wrongly. How could we ever tell?

We could guard against such contingencies by building two identical clocks. Provided they both read the same time, could we not rest assured that the time was right? It sounds appealing, but it is a basic principle of timekeeping that two clocks will *never* keep the same time. We may construct two clocks, identical as far as we can make them, but no matter how closely we set them their readings will in due course start to diverge. This is simply because at some level the rate of the two clocks will differ, even by a minuscule amount, and as time goes by that difference can only grow.

It gets worse. If we had two clocks and they read differently, how would we know which was in error? There is no way to decide which of two identical clocks is "right". Three clocks are better, because if one is in error the other two will still agree (within some limit) and we'll know which one is faulty. But perhaps we should make that four clocks, in case one breaks down. On the other hand, five clocks would ensure that one could safely be taken out of service for regular maintenance ....

By now you may have guessed that there is no world clock. There is no clock in Paris pumping out SI seconds, but there are two time scales that emanate from BIPM. They are called International Atomic Time (TAI) and Coordinated Universal Time (UTC).

World time scales

As we have seen, the 1960s were a time of confusion, as authorities attempted to reconcile the precise and uniform time scale provided by atomic clocks with the fluctuating rotation of the Earth and the steady but inaccessible scale of Ephemeris Time. But from 1967, with the second finally shorn of its astronomical heritage and defined uniquely in terms of the caesium transition, the way was open to resolving the international muddle over time scales. In 1971 the CGPM designated the BIH atomic time scale as TAI (Temps Atomique International). At the same time the International Radio Consultative Committee—part of the International Telecommunication Union (ITU)—recommended a new form of Coordinated Universal Time (UTC), closely linked to TAI, that was to be used for time signals. A new system of timekeeping based on TAI and UTC was introduced worldwide from 1 January 1972.

The timekeeping system we have today is essentially unchanged since then, apart from a significant administrative milestone. The BIH, which had coordinated timekeeping since 1920, was wound up by the International Astronomical Union on 1 January 1988. Its timekeeping functions were transferred to the newly created Time Section of the BIPM at Sèvres, while its astronomical responsibilities were assumed by a new International Earth Rotation Service (IERS), with its central bureau at the nearby Paris Observatory. We will find out more about the work of the IERS in the next chapter.

TAI is the continuation of the BIH's atomic time scale whose origins can be traced back to the NPL caesium standard in 1955. It is a uniform scale whose interval is the SI second as defined in 1967 in terms of the caesium transition frequency. Today it is formed not from a world clock in Paris, but from the combined readings of more than 260 atomic clocks around the world calibrated by less than a dozen primary frequency standards.

UTC is the basis for all civil timekeeping, occupying the same position as GMT did until the 1920s. When you hear radio time signals you are hearing UTC, shifted, perhaps, by a whole number of hours to match your time zone. It, too, counts precise SI seconds and differs from TAI only by a whole number of seconds. It is kept closely in line with the rotation of the Earth. Whenever changes in the length of the day threaten to cause UTC to depart by more than 0.9 seconds from UT1—which, as you may remember, is UT defined by the orientation of the Earth—a "leap second" may be introduced to bring it back into line. TAI and UTC were synchronised on 1 January 1958 and by the beginning of the year 2000, UTC lagged precisely 32 seconds behind TAI.

We shall have more to say about how TAI and UTC are formed, but first we shall look at the successors to the NPL atomic clock, the primary frequency standards which realise the duration of the second.

Primary standards

At the heart of the world's timekeeping system are the primary frequency standards, every one of them based on the caesium transition. Their job is to measure off precise SI seconds without reference to external comparisons. There are precious few primary standards. The BIPM annual report shows that in the six years to 1998 only 11 primary standards in seven laboratories were available to contribute to TAI and not all of them are in regular operation (Table 4.1). On occasion, as few as two or three may be in service. These machines are not intended to tell the time (though a few do), but simply to measure the length of the second. Each of them is unique, built by hand to attain the most exacting performance achievable by modern physics. Each of them is, to some extent, experimental and that is why, with a few exceptions, they do not run continuously. Their designers are forever improving them, tweaking them here and there, coaxing just a little more accuracy out of them.

Their accuracy in measuring the second varies from about one part in 10^{13} to two parts in 10^{15}, about a factor of 50. Thomas Harrison's chronometer, which revolutionised navigation in the eighteenth century, was required to be accurate to 3 seconds a day: the best of these frequency standards have an accuracy exceeding 1 nanosecond a day. The duration of the second is known to 14 decimal places—by far the most accurately determined of any of the physical units of measurement.

Classical beams

The best of the classical caesium beam standards, similar in principle to the NPL standard of the 1950s, are widely reckoned to be the machines built and operated by the Physikalisch-Technische Bundesanstalt (PTB) at Braunschweig in Germany. PTB is the national standards laboratory for Germany, and it was here in the 1930s that scientists first discovered the seasonal variations in the rotation of the Earth using quartz-crystal clocks.

Table 4.1. The world's primary frequency standards contributing to TAI (1993–98).

Institution	Location	Name	Uncertainty (parts in 10^{14})	Type
Communications Research Laboratory	Tokyo, Japan	CRL-O1	1.0	Optically pumped caesium beam
Institute of Metrology for Time and Space	Mendeleevo, Russia	SU MCsR 102	5	Caesium beam
Laboratoire Primaire du Temps et des Fréquences	Paris, France	LPTF-JPO	11	Optically pumped caesium beam
		LPTF-FO1	0.22	Caesium fountain
National Institute of Standards and Technology	Boulder, USA	NIST-7	0.7–1.0	Optically pumped caesium beam
National Research Council of Canada	Ottawa, Canada	CsVI A	10	Caesium beam
		CsVI C	10	Caesium beam
National Research Laboratory of Metrology	Tsukuba, Japan	NRLM-4	2.9	Optically pumped caesium beam
Physikalisch-Technische Bundesanstalt	Braunschweig, Germany	CS1	0.7–3.0	Caesium beam
		CS2	1.5	Caesium beam
		CS3	1.4	Caesium beam

Figure 4.1. Chief technician Harald Brand with CS2, one of the four caesium beam standards operated by the German standards laboratory, PTB. CS2 runs continuously as a clock, providing UTC for Germany. For nine years until 1995 it was the most accurate clock in the world. Brand was responsible for most of the construction of CS2 and has recently retired after more than 30 years with PTB.

The first of the PTB primary standards, CS1, came into operation in 1969. It ran intermittently until 1978 and then continuously until 1995. After refurbishment it has been operating continuously again since 1997. Its successor, CS2, has been running continuously since 1986 (Figure 4.1). For nine years until 1995 CS2 was the most accurate primary standard in the world. CS1 and CS2 were innovative in that the deflecting magnets were designed to keep the flipped atoms tightly focused along the beam axis, reducing errors associated with large deflections. Atom speeds are below 100 metres per second, less than half that in the original NPL design. Both machines are designed to be used in either direction— there is an oven and a detector at each end of the tube—and this helps to eliminate other sources of error.

Two more recent standards, CS3 and CS4, started up in 1988 and 1992. In these the atoms emerge even more slowly into the beam tube— about 70 metres per second—and for that reason the tubes are mounted vertically. If they were mounted horizontally, like their cousins, the atoms would drop noticeably in their flight from one end to the other. Neither has so far proved as reliable as CS2, and CS4 has yet to contribute to TAI.

CS1 and CS2 are operated continuously as clocks as well as frequency standards. Indeed, CS2 provides UTC to Germany as well as being the most reliable contributor to TAI.

Optical pumping

The PTB clocks' long reign as the world's best primary standards came to an end in 1995 when the latest standard from the US National Institute for Standards and Technology (NIST, formerly the National Bureau of Standards) became operational. NIST-7 (Figure 4.2) is the seventh in a line of standards reaching back to NBS-1, which is how the first NBS caesium beam machine was designated after it was reassembled in Boulder, Colorado, in 1958.

While working on the same basic principles as the PTB clocks, it makes much more efficient use of the caesium atoms. In a classical caesium standard the atoms are separated by a magnetic filter according to the hyperfine state they happen to be in. Instead of magnetic filters, NIST-7 uses a method known as optical state selection (Figure 4.3). The caesium atoms emerging from the oven pass through a laser beam whose photons have just the right energy to lift the outer electron from the lower hyperfine state into a higher energy level, but leave those in the upper state untouched. After a few nanoseconds the electron drops down again, but is equally likely to fall into either of the hyperfine states; that is to say, with its magnetic field pointing either up or down. Since the

Figure 4.2. NIST-7, one of the most accurate atomic clocks in the world, is pictured here with its developers John P. Lowe, Robert E. Drullinger (project leader), David J. Glaze, Jon Shirley and David Lee of the US National Institute of Standards and Technology. It is an optically pumped caesium beam standard which is accurate to 2 seconds in 10 million years.

laser continually removes electrons from the lower state and returns them equally to both, a short flash from the laser quickly "pumps" all the atoms into the upper state. Now they can pass into the Ramsey cavity.

As usual, if the microwaves are near the caesium frequency the atoms flip over and on emerging from the cavity they are again in a mixture of states: those still in the upper state and those that have been knocked down to the lower state. Another laser selectively excites the atoms remaining in the lower level, and as they fall back the photons they emit are detected. The more photons are detected, the more atoms are in the lower state, and the closer the microwave radiation is to the transition frequency.

The beauty of optical state selection is that all the caesium atoms

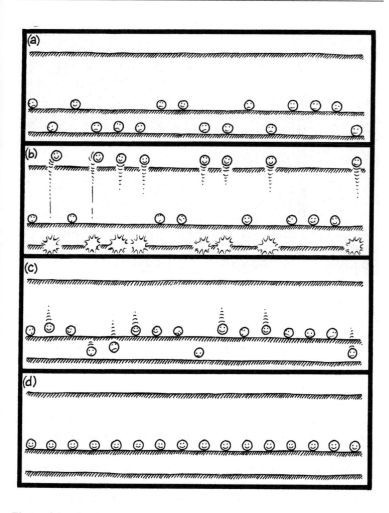

Figure 4.3. Optical state selection of atoms. (a) In a bunch of normal caesium atoms, about half the electrons will be in the lower hyperfine state and half in the upper state. (b) When illuminated with a laser of the right wavelength, the electrons in the lower state are excited to a higher level. (c) The excited electrons soon drop back, about half falling into each of the upper and lower states. (d) After a few cycles all the electrons are in the upper state and ready to be exposed to microwaves.

are used and none is wasted. Because there are no magnetic filters the atoms fly in a straight line from one end of the machine to the other and this reduces other sources of error. This means that NIST-7 can detect the flipped atoms much more reliably than conventional beam standards, which means that higher accuracy can be attained. Three more optically pumped primary standards, one in France and two in Japan, have followed in the pioneering footsteps of NIST-7.

The caesium fountain

We have already met Jerrold Zacharias, the MIT physicist who astonished the world by designing a commercial caesium clock which was on the market little more than a year after the NPL machine started working. A few years earlier Zacharias had had an even more radical idea for an atomic clock. We have seen that the performance of a caesium clock depends on the time it takes the atoms to pass between the two arms of the Ramsey cavity—the longer the time between the two exposures to microwaves the more stable the frequency generated by the clock. It occurred to Zacharias that instead of shooting the atoms through a horizontal cavity at 200 metres per second, it should be possible to launch them vertically, like a fountain.

The idea was that the atoms would be propelled upwards from the oven, be slowed by gravity to a momentary standstill, then fall back down again. They would be exposed once to microwaves on the way up and again on the way down (Figure 4.4). If only the slowest atoms emerging from the oven could be used then they would spend as long as a second between exposures—about 200 times longer than was possible with the caesium beam machines then being developed. In 1953 Zacharias started building a 5-metre high machine which he called the Fallotron. Unfortunately it never worked. It turned out that the slow atoms essential to its operation were being knocked out of the beam by the faster atoms.

It was not until the 1990s that new discoveries in physics made it possible to create a practical frequency standard based on the Zacharias idea. Since 1995 a caesium fountain has been in operation at the Laboratoire Primaire du Temps et des Fréquences (LPTF) at the Paris Observatory (Figure 4.5). Clouds of a million caesium atoms are

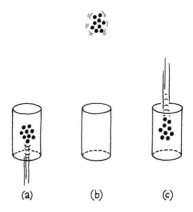

Figure 4.4. Principle of a caesium fountain. (a) A cloud of cold caesium atoms is propelled up through a cavity where they are exposed to microwaves. (b) The atoms reach a momentary standstill before falling back down. (c) On their second pass through the cavity the atoms receive another exposure to the microwaves. The interaction time—the interval between the two exposures—is much greater than for a caesium beam standard and so the transition frequency can be defined more precisely.

launched every second. LPTF-FO1 can determine the length of the second to an unprecedented accuracy of two parts in 10^{15} and has already contributed to TAI.

A second fountain, NIST-F1, made its first contribution to TAI at the end of 1999. So promising are caesium fountains that no one is building any more classical caesium beam machines. The future is in fountains, as we shall see in Chapter 8.

Secondary clocks

If we had to rely on this handful of primary standards the world's time-keeping system would be precarious indeed. Few of these machines operate all the time. How, then, can we keep track of atomic time in the periods when the primary standards are out of service?

Figure 4.5. The first operational caesium fountain, LPTF-FO1, has been contributing to the world timekeeping system since 1995. Based at the Paris Observatory, the machine is currently the most accurate frequency standard in the world and can keep time to better than 1 second in 10 million years.

TAI is kept going by the combined efforts of an "ensemble" of 260 secondary atomic clocks dispersed among the world's timekeeping laboratories (Figure 4.6). Not only does the ensemble guard against losing count of atomic time; when properly used, it can provide a time scale more stable than that of even the best clock in the group. There are two kinds of ensemble clocks: industrial caesium clocks and hydrogen masers.

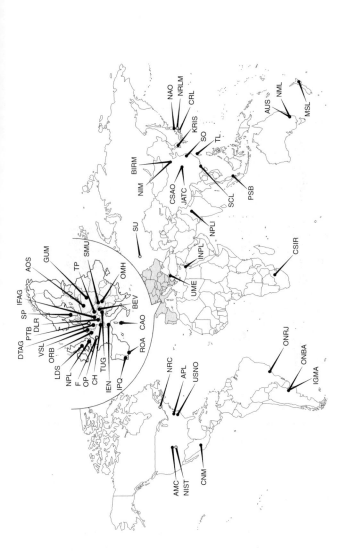

Figure 4.6. UTC is formed by the collective efforts of about 50 timing centres around the world, each of which maintains one or more atomic clocks. The seven laboratories operating primary standards are marked with an open circle.

Industrial caesium clocks

By far the majority of the world's 260 atomic clocks contributing to TAI are industrial caesium standards. These are similar in principle to the primary standards, but manufactured commercially and can be bought off-the-shelf. They are designed to be self-contained and pretty much foolproof. When the clock runs out of caesium the whole of the beam tube is replaced with a new one.

While the primary standards are built with beam tubes a metre or more in length, industrial standards are necessarily more compact. Their tubes can accommodate a Ramsey cavity no more than 15 centimetres long and this limits their accuracy. The oven and detector are angled away from the main axis of the tube so that only one of the hyperfine states passes into the cavity (Figure 4.7). This reduces the proportion of unflipped atoms that reach the detector and also removes the faster atoms before they enter the cavity. Before reaching the detector the beam of flipped atoms may be passed through a device which filters out any contaminant atoms. All these features help to compensate for the short tube.

In recent years a new type of caesium clock, the HP 5071A made by Hewlett-Packard, has become the instrument of choice for the timing centres. First appearing on New Year's Eve in 1991, the HP 5071A has an improved design of beam tube and electronic control which makes it ten times more accurate than previous clocks and ten times less sensitive to environmental effects. Its long-term stability, usually better than one part in 10^{14} over a period of 40 days, is such that it will keep time to better than a second in 1.6 million years. The HP 5071A is so popular that it now accounts for almost half of the industrial clocks contributing to TAI and carries over 80 percent of the weighting. As a result the stability of TAI and UTC have improved by an order of magnitude in the 1990s.

Hydrogen masers

In the early 1960s when caesium clocks were revolutionising timekeeping, another type of atomic clock looked very promising. We have already seen in Chapter 2 how the hydrogen atom, with the hyperfine splitting of its ground state, has a "clock" transition at 1420 megahertz.

Industrial Caesium Clock

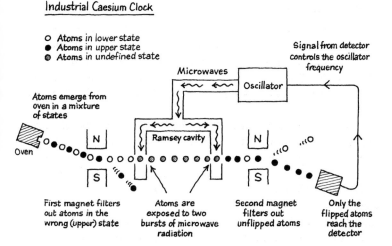

Figure 4.7. Industrial caesium clocks are more compact and robust than primary caesium standards (Figure 2.13) but similar in principle. The main difference is that the oven and detector are offset from the axis of the beam tube. The Ramsey cavity may be only 15 centimetres long.

In 1960 a group at Harvard University led by Norman Ramsey (the man who invented the "Ramsey cavity" used in caesium beam standards) developed the first hydrogen maser.

Hydrogen gas is heated in an electric discharge lamp until the molecules split into atoms. These atoms are much like caesium atoms in that they are divided more or less evenly between the two hyperfine states. As in an industrial caesium standard, a beam of hydrogen atoms passes through a magnet which permits only atoms in the upper state to pass into the rest of the device. Instead of shooting through a cavity, the atoms are caught in a quartz bulb coated on the inside with Teflon, the material used for making non-stick frying pans. Thanks to the Teflon, the atoms do not stick to the wall (this is not a joke!) but bounce around inside the bulb for about a second or so. During this time they are exposed to microwaves near the transition frequency.

When the radio frequency corresponds exactly to the transition frequency, the electron is knocked down to the lower state and the atom emits a photon of exactly the same frequency. Under the right conditions these emitted microwaves can be detected by a receiver and used directly to stabilise a quartz oscillator.

This type of hydrogen maser is superior to caesium clocks over periods from a few seconds to a few days, attaining a stability of one part in 10^{15}. This makes it the standard of choice for demanding applications such as radio astronomy (see Chapter 7) but it is not as good as caesium for longer term stability. Nonetheless, about 50 hydrogen masers around the world contribute towards TAI, and they are especially popular in Russia. None are used as primary standards. NPL has two, one of which is the clock providing UTC for the United Kingdom.

Rubidium clocks

There is one other kind of atomic clock that is widely used, although it does not contribute to TAI. Shortly after spectroscopists Bunsen and Kirchhoff discovered caesium in 1860, they stumbled across a second new metal. This one had a pair of bright red spectral lines very similar to the yellow lines of sodium, and was named "rubidium" after their ruby-like glow. Rubidium, it turned out, was another of the alkali metals, similar to caesium, and possessing the single outer electron that makes caesium so suitable for atomic timekeeping. Like caesium, its ground state is split into two hyperfine states, but unlike caesium, rubidium occurs naturally in a mixture of two isotopes. About 72 percent occurs as rubidium-85, which has a transition frequency of 3035 megahertz, while 28 percent is found as rubidium-87, with a frequency of 6835 megahertz. The difference is only two neutrons, but that makes a big difference to the magnetism of the nucleus, which in turn affects the hyperfine splitting.

Rubidium clocks do not work in the same way as caesium clocks. A lamp containing hot rubidium-87 vapour emits light containing the twin red lines at 780 and 795 nanometres (Figure 4.8). The light passes through a cell containing cooler rubidium-85 vapour, but because of the difference in the hyperfine structure of the ground states of the two isotopes, the light emerges with a subtle modification: if it is now passed through a cell of rubidium-87 vapour, the light will be absorbed by atoms

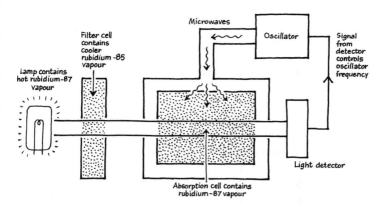

Figure 4.8. The cheap and cheerful rubidium standard can be used as an atomic clock where the utmost accuracy is not required. The absorption cell darkens when the microwaves are tuned to the transition frequency of 6835 megahertz (see text for details).

in the lower hyperfine state but not by those in the upper state. The result is rather similar to the optical pumping in NIST-7, in that rubidium-87 atoms all end up in the upper hyperfine state. When no atoms are left in the lower hyperfine state the gas can no longer absorb the light and becomes transparent. If the gas is now exposed to microwaves near the transition frequency of 6835 megahertz the electrons will be knocked down to the lower hyperfine state. The atoms will once again be able to absorb light and so the cell goes dark. So the idea is to adjust the radio frequency from a quartz-crystal oscillator until the cell goes dark, and you then know that the oscillator frequency is the same as the transition frequency.

Although it cannot compare in performance to the caesium clock— it suffers from relatively poor accuracy and ageing effects—the great advantage of the rubidium clock is its compact size and its low cost. It is more stable than quartz-crystal clocks, attaining a stability of one part in 10^{13} over 1 day. Specially stabilised rubidium clocks are used in the satellites of the US Global Positioning System (GPS). We shall have a lot more to say about GPS in Chapter 6.

Making the time

How do all these primary standards and atomic clocks combine to give us TAI and UTC? In Chapter 2 we saw that any clock can be thought of as containing two parts: an oscillator and a counter. Although it exists only on paper, the TAI "clock" can be regarded in a similar fashion: the primary standards provide the oscillator and the ensemble of clocks in national timing centres is the counter. The ensemble also acts as a flywheel, keeping the TAI clock going even when the oscillator is not available. It "remembers" the periodic calibrations from the primary standards.

The job of producing and disseminating a time scale from this widely dispersed set of clocks falls upon the staff of the BIPM Time Section. They have it down to a fine art. The first step is to gather timing information from the ensemble of 260 clocks operated by the national centres. NPL, for example, maintains six atomic clocks, one of which provides UTC for the United Kingdom (Figure 4.9). This local version of UTC is designated UTC(NPL). Once an hour the clocks are automatically compared with each other and their differences recorded to the nearest nanosecond. At the end of each month NPL and the other 50-odd centres compile their measurements at five-day intervals and send them in a standard report by e-mail to BIPM.

This information tells BIPM how the clocks within each centre are performing in comparison with local UTC, but by itself is not enough to compare the different versions of UTC. How does UTC at NPL compare with, say, UTC as produced by PTB in Germany or by NIST in the US?

That is done by means of the GPS, the network of navigation satellites maintained by the US Department of Defense. Each satellite broadcasts time signals from its own internal atomic clock. Thirty times a day the GPS receiver at NPL automatically tracks a satellite, comparing its time signals with UTC(NPL). All around the world other timing centres are doing the same, according to a schedule issued by BIPM and designed to ensure that pairs of centres are observing satellites at the same time. By a complex series of processes BIPM is able to calculate the difference between each local UTC and UTC(OP); the UTC at the nearby Paris Observatory.

At the end of each month the staff at BIPM consider the previous 30 days of timing data. For five-day intervals they know how the clocks in

Figure 4.9. NPL scientist Andrew Lowe at work in the laboratory's timing centre. Two of NPL's six atomic clocks (with the open flaps) are visible in the lower part of the picture. The unit in the centre of the picture (with the NPL logo) is displaying UTC(NPL)—UTC for the United Kingdom—which is derived from a hydrogen maser clock in the basement. Other equipment is used to monitor time signals from standard frequency radio stations and to receive signals from GPS satellites.

each timing centre differ from the local UTC, and they know how each local UTC differs from the UTC(OP). Simple arithmetic gives a version of UTC for every one of the 260 clocks in the ensemble.

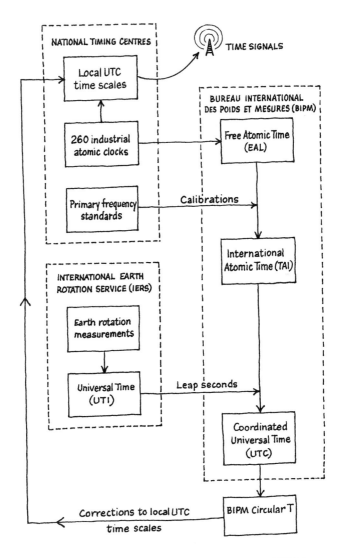

Figure 4.10. The formation of Coordinated Universal Time (UTC) is a complex process involving dozens of organisations in many countries.

The next step is to take a kind of average of these times to create an "average" clock. The time scale generated by this average clock is known as Free Atomic Time (Echelle Atomique Libre, EAL). To form the average, each of the contributing clocks is given a "weight" according to how stable it has been in comparison to EAL over the previous year. This weighting is given according to how well the clock has performed in practice, rather than on the reputation of the model or its manufacturer. A clock with a proven record of keeping close to EAL is given more weight than a clock which has not been performing so well or which has been newly introduced. The most reliable clocks are allowed to contribute up to 0.7 percent towards EAL, while any clock which is seen to be behaving abnormally can be given zero weight—that is, it is ignored. The weighting formula ensures that the best clocks count most to the stability of EAL without a small number of exceptionally good clocks being allowed to dominate.

Next, the scale interval of EAL is compared with the length of the SI second generated by the primary standards. Several times a year the standards transmit their measurements of the second to BIPM. At the end of the 1990s the interval of EAL was shorter than the SI second by about seven parts in 10^{13}. When the interval of EAL is adjusted accordingly, it becomes TAI. At present, the TAI second is believed to correspond to the SI second to within four parts in 10^{15}.

The final step is to convert TAI to UTC. As we have seen, UTC and TAI differ by a whole number of seconds and are defined so that the two coincided on 1 January 1958. An extra second—a leap second—is occasionally introduced into UTC to keep it within 0.9 seconds of UT1, the time kept by the rotating Earth. The decision whether or not a leap second is required is taken by the International Earth Rotation Service, and we will be looking at that process in the next chapter. After the leap second on 31 December 1998, UTC was running 32 seconds behind TAI. Figure 4.10 summarises the whole process.

Coordinated Universal Time

Civil time around the world is based on UTC emanating from one of the 49 national timing centres. In the UK it is UTC(NPL), in France

UTC(OP) while in Germany it is UTC(PTB). In the US two versions of UTC are available, one from NIST and the other from USNO. When you hear time signals on the radio or telephone, they can be traced to one or other of the local sources of UTC.

Most laboratories designate one of their clocks to provide their version of UTC. A few generate a "paper" UTC by averaging a number of clocks in much the same way as BIPM, and this can give a very stable time scale. Each of these local versions of UTC will differ by a small amount from UTC disseminated from Paris. These differences are collated in a monthly bulletin called Circular T, issued by BIPM. Its six pages of tables are essentially a report to the national timing centres on how well they have done in generating their local UTC. It can be conveniently accessed from the BIPM website.

As an example, we'll look at Circular T issued on 20 December 1999. This is based on data collected at five-day intervals between 28 October and 27 November. The Circular tells us the difference between each local UTC and UTC disseminated from BIPM. In that period UTC for the United Kingdom was running 60–80 nanoseconds behind UTC. The most accurate UTCs were being maintained in the Netherlands (within 10 nanoseconds), the US (11) and Sweden (18). Even the largest deviations were only a few microseconds, which is good enough for most purposes. With this information, users who require the most accurate time can correct their measurements to UTC according to which local version they were using. Circular T also reports that during that period calibrations were received from the primary standards LPTF-JPO, NIST-7, NRLM-4 and CS1, CS2 and CS3 at PTB. It also reveals how time disseminated from navigational satellites compares with UTC, and how 16 local atomic time scales compare to TAI. Information from the Circular Ts is compiled into the annual report of the BIPM Time Section which is published every February.

It is a curiosity of the world timekeeping system that only local approximations of UTC are available in "real time". We have to wait a month or more to know what UTC was to its full accuracy. The question arises as to what a timing centre can then do when they find their own UTC is a long way from UTC announced in Circular T. Stability is all important. If the local time is maintaining a more or less constant

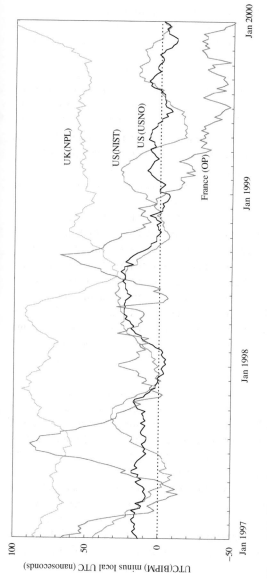

Figure 4.11. Examples of well-performing local UTCs. The national timing centres are encouraged to keep their local versions of UTC within 100 nanoseconds of UTC computed by BIPM. USNO always does well because its many atomic clocks contribute heavily in the averaging process. Sudden changes in the tracks, such as OP (Paris Observatory) in summer 1997, are caused by the rate of the local UTC being slightly adjusted to keep it near to UTC.

difference to UTC, then no action need be taken. On the other hand, if the difference is changing rapidly, then the clock must be adjusted before local time departs too far from UTC.

The process is aptly known as "steering" and is a delicate operation, much like steering an oil tanker. For one thing, any deviation of local time from UTC is only discovered when Circular T appears, which may be several weeks after the event. A steering correction applied then will take immediate effect, but the laboratory will not know the outcome for several more weeks. The everyday procedures for adjusting clocks do not apply. You or I might correct a domestic clock by moving the hands to the correct time, but while such drastic measures are occasionally used to correct atomic clocks, the laboratory usually brings its time closer to UTC in a more gradual fashion. The preferred method is to alter the rate of the clock by a small amount known as a "frequency step". Suppose we had a clock running at a steady 100 nanoseconds ahead of UTC and wished to reduce that difference. This could be done by applying a frequency step of minus 1 nanosecond per day for 100 days. In that time the clock will approach closer and closer to UTC. After 100 days, when the clock reads close to UTC (which of course can only be confirmed several weeks later), a frequency step of plus 1 nanosecond per day should ensure that it stays there. Of course, things are never that simple, and even some of the leading laboratories have to constantly steer their clocks this way and that to keep them close to UTC. Figure 4.11 shows the performance of a few selected timing centres over an extended period.

Local deviations from UTC have reduced dramatically in recent years. In the 1980s even the leading centres could be several microseconds in error, but now the best centres are routinely within a few tens of nanoseconds of UTC. Partly this is because timing centres have invested in better clocks, especially the HP 5071A and hydrogen masers, and maintained them in environmentally controlled conditions. But they have also been more active in steering their local UTCs closer to UTC. BIPM now requests that all centres keep their UTC within 1 microsecond of UTC and preferably within 100 nanoseconds. Most of them manage the former but not many achieve the latter.

The strange fate of GMT

In 1975 the CGPM recommended that UTC become the basis for legal timekeeping in all countries, replacing Greenwich Mean Time which now had no clear meaning. While this sensible recommendation has been enacted in most industrialised countries it was not adopted in the United Kingdom.

Legal time in the UK remains GMT, even though GMT was abolished for scientific purposes in the 1920s and replaced by Universal Time, which itself was subdivided into UT0, UT1 and UT2 in the 1950s. Quite what this means for GMT no one knows. If GMT was originally defined as mean solar time determined from observations at Greenwich, then it could now be interpreted as UT0, the time at a single observatory. If it is regarded as the basis of world time, as it was from 1884, then UT1 would seem appropriate since the astronomers of the day would surely have corrected GMT for polar motion if they had known about it. One could even make a case that UT2, which is UT1 with seasonal fluctuations smoothed away, is closest to the uniform time scale which GMT was understood to represent when it was introduced. It is certainly not identical with UTC, since GMT was not derived from atomic time and had no leap seconds. The best one can say is that, if GMT still exists in some limbo reserved for undead time scales, then it could differ by up to a second from UTC.

What is certain is that no clocks read GMT. BBC and telephone time signals in the UK and worldwide have broadcast UTC since 1972, and when the British set their clocks they set them to UTC. Even Big Ben chimes UTC, as near as it can manage. This does not appear to have discouraged people from believing that their clocks show GMT, at least in the winter.

In order to correct this anomaly, the Coordinated Universal Time Bill was introduced in the House of Lords in June 1997. Its sponsor, Lord Tanlaw, pointed out that for all intents and purposes time in the UK was based on UTC and not the legal but undefined GMT, which could differ from UTC by anything up to 1 second. While such differences were small, it seemed unreasonable to have a legal time that differed from clock time and fluctuated according to unforeseeable changes in the rotation of the Earth. And there could be problems with legal documents when timing was critical, especially with computers and other devices increasingly taking their time direct from UTC sources. Tanlaw's bill proposed simply to replace all legal references to GMT with UTC. Mindful of the new government's enthusiasm for celebrating the millennium in style he added: "When the eyes of the world are focused on Greenwich and its history at the start of the new millennium, it would be a very sad event indeed if we, as a nation, failed to display the correct time when the moment came."

Tanlaw's bill was approved by the Lords on 14 July 1997 and passed to the House of Commons. And what happened to it there? Why, it ran out of time.

5

THE LEAP SECOND

International Earth Rotation Service

It may seem strange to devote a whole chapter to leap seconds, but the story of how extra seconds are occasionally introduced into UTC is a long and complex one. We'll begin with an organisation called the International Earth Rotation Service (IERS), whose sole job is to monitor the rotation of the Earth and its orientation in space.

The IERS came into being on 1 January 1988. It took over the astronomical functions of the BIH, whose responsibility for atomic time was transferred to BIPM at the same time. IERS also took over the former International Polar Motion Service, which was itself descended from the International Latitude Service which started continuous monitoring of polar motion in 1899.

IERS has many responsibilities apart from its duties in timekeeping. First, it maintains the International Terrestrial Reference Frame, which defines the precise locations of several hundred reference points on the surface of the Earth to within a centimetre or so. This has many applications in geology and geophysics. Secondly, IERS maintains the International Celestial Reference Frame, which is a positioning system widely used by astronomers and based on the precise positions of more than 500 distant galaxies. The two frames are connected by continuous monitoring of the rotation of the Earth.

Five quantities are routinely measured and between them they completely specify how the Earth is oriented in space. Four concern the direction of the rotation axis: two of them specify the position of the pole on the ground in comparison with a standard reference pole and the other two specify the direction the pole is pointing in space. The fifth quantity, and the most useful for timekeeping, is the type of Universal Time known as UT1.

In Chapter 3 we saw that, since the 1950s, Universal Time has been classified into UT0, UT1 and UT2. If a single observatory measures UT from astronomical observations that time is UT0. With observations from several observatories it is possible to correct for polar wobble and the resulting UT is called UT1. UT1 is essentially a measure of the rotational orientation of the Earth in space—the angle it has rotated through. It is the time that is of direct interest to navigators who use the Sun and stars, though it has to be said that there are not many of those these days.

A third type of UT—known as UT2—is derived from UT1 by applying a standard formula which corrects approximately for the seasonal fluctuations in the rotation rate. As a smoothly running time scale UT2 was important in the 1950s and 1960s but since the redefinition of UTC in 1972 it no longer features much in timekeeping.

Between them, BIPM and IERS have to ensure that UTC stays at all times within 0.9 seconds of UT1. If it appears that UTC is likely to go beyond that limit then IERS instructs BIPM to insert a leap second. This chapter is concerned with how IERS monitors the rotation of the Earth to obtain UT1, and how it decides when a leap second is required.

How UT1 is measured

All methods rely on determining the orientation of the Earth with respect to external objects such as galaxies, stars, the Moon and artificial satellites.

Traditional methods

Until the 1980s measurement of Universal Time relied on measuring the time at which the rotating Earth carried a known star across the meridian. What was being measured was really sidereal time (see Chapter 1) which was then corrected to give Universal Time, or strictly UT0. By gathering observations from a network of stations several values of UT0 could be combined to give UT1.

The simplest method is based on a specialised telescope called a transit circle, first used by Danish astronomer Ole Römer in 1689 (Figure 5.1). It is a refracting telescope mounted so that it can sweep precisely along the meridian from north to south but not away from it. The method relies on knowing the precise position on the sky of a number of standard stars. Let us suppose we use the bright star Capella. From astronomical tables we can find that on 1 January 2000 Capella will cross the meridian at 05:16:42 local sidereal time, which we can convert to UT0 using standard formulae. So we point the transit circle at the meridian ahead of Capella and wait for it to cross. The instant it crosses the line we know UT0 and can set our clock accordingly. Using automated equipment it is possible to measure UT to within a few milliseconds with a transit circle.

A more sophisticated device is the photographic zenith tube (PZT), a fixed telescope that points straight upwards. Designed in 1909, the PZT was, until very recently, the instrument of choice for precise monitoring of the Earth's rotation. An image of a small patch of sky directly overhead is reflected from a pool of mercury (defining a precise horizontal surface) and four images of the stars are recorded on a moving photographic plate along with the clock times of the exposures. Careful measurement of the plate gives a very accurate transit time. Because it is precisely aligned on the zenith, the PZT can measure the latitude of the observatory as well as UT, and so monitor polar motion. PZTs were used by the US Naval Observatory to relate UT to ET as part of the calibration of the atomic second in the 1950s.

The newest device for measuring Earth rotation by the stars is the Danjon astrolabe, invented by the same André Danjon who first proposed the use of Ephemeris Time in the 1920s. First used in 1951, the astrolabe determines the instant at which a star reaches an altitude above the horizon of precisely 30 degrees. It can look in any direction, so can measure many more stars than either a transit circle or a PZT.

Ultimately, each of these techniques is limited by the accuracy with which the instrument can be aligned and the positions of stars listed in catalogues. What really made optical methods obsolete was the much greater precision that became possible from other techniques in the 1970s and 1980s. Indeed, a few modern transit telescopes are still in use, but they are now used the opposite way: knowing the precise time of transit from an atomic clock, astronomers can get accurate positions of the stars.

Figure 5.1. Until the late twentieth century all our timekeeping was based on the timing of stars as they crossed the meridian. This specialised telescope, the Airy Transit Circle, kept time at the Royal Observatory in Greenwich from 1851 to 1927, when it was replaced with more modern instruments. The object lens of the telescope is the grey circle in the centre of the picture. In 1884 the location of the telescope was internationally agreed to define the prime meridian of longitude.

VLBI

The most powerful of the modern methods of monitoring the Earth's rotation is called very long baseline interferometry, or VLBI. Developed from techniques pioneered in radio astronomy in the 1950s, VLBI links together many widely spaced radio telescopes, often thousands of kilometres apart, so that they act in concert to "synthesise" a much bigger telescope equal in size to the widest spacing between the observatories. We will not go into the detail here, save to say that each observatory agrees to observe the same astronomical object at the same time and records the signals on a high-speed tape deck. The tapes are then played back together and the signals combined to create an image of the object in fabulous detail. VLBI observations now produce radio images which are far more detailed than those from optical telescopes. A consequence of this is that the positions of some of the most distant objects in the universe—the fantastically powerful galaxies known as quasars—are known far more accurately than any star. It then becomes possible to use this precision to determine UT1.

To use VLBI for Earth rotation studies, the times at which radiowaves from a known quasar arrive at each of the telescopes are extracted from the tapes. The differences in arrival times can then be used to calculate the orientation of the Earth with respect to the quasar. Because the position of the quasar is accurately known, the orientation of the telescope array, and hence the Earth, can be calculated to less than one-thousandth of a second of arc.

VLBI observations are routinely carried out on about 500 quasars. In 1997 more than a quarter of a million observations were made from 35 telescopes. Among these are regular observations carried out by networks in the US, Europe and Japan.

VLBI measures the rotation of the Earth in comparison to the most distant objects in the universe. Unless the universe as a whole is rotating in some sense, then this technique determines the absolute orientation of the Earth in space, both the position of the pole and UT1.

Lunar laser ranging

When Apollo 11 landed on the Moon in July 1969, one of the packages placed on the Sea of Tranquillity by astronauts Neil Armstrong and

Edwin Aldrin was a flat box containing an array of 100 light reflectors. Similar in principle to the reflectors on the backs of cars, the array was designed to reflect pulses of laser light from Earth straight back to where they came from. The idea was to measure the precise distance from the Earth to the Moon.

Light travels through space at almost exactly 300 000 kilometres per second. The Moon's distance varies, but on average it is about 384 000 kilometres from Earth. A pulse of light will take about 1.3 seconds to get from the Earth to the Moon. The round trip, after being reflected from the Apollo 11 array, will be about 2.6 seconds. If the time between sending the pulse and seeing the reflected flash can be measured accurately, then the distance to the Moon can be calculated with equal accuracy.

Following the Apollo 11 mission, a similar array was left in the Fra Mauro uplands by Apollo 14 in 1971 (Figure 5.2) and a bigger one in the Apennine foothills by Apollo 15 in the same year. Arrays were also taken to the Moon on the Soviet automated rovers Lunokhod 1 (Sea of Rains 1970) and Lunokhod 2 (Sea of Serenity 1973), although the former is no longer operating.

Lunar laser ranging presents formidable challenges. Normal sources of light are not strong enough to do the job and spread out too much: even a pulse from a powerful laser spreads to a circle about 7 kilometres in diameter at the distance of the Moon. A tiny fraction of that light—about a billionth—falls on the array and is reflected back. The returning flash from the Moon also spreads out, and forms a circle 20 kilometres in diameter on the Earth. A collecting telescope picks up at most only two-billionths of that light. Add to that the other losses, and it turns out that out of every 10^{21} photons sent towards the Moon only one will be detected back at the Earth.

Small wonder that very few observatories have the dedication to continue lunar laser ranging. The main centres since 1969 have been the McDonald Observatory in Texas, the Haleakala satellite station on Maui in Hawaii (now discontinued), and a purpose-built station at Grasse in the south of France. Interest does seem to be picking up again, with a new station being constructed at Matera in southern Italy.

The accuracy of lunar ranging has improved steadily since the 1970s, and it is now possible to measure the distance between the ground

Figure 5.2. One of the four laser reflectors still operating on the Moon and used regularly for Earth rotation studies. This one was left in the Fra Mauro region by the crew of Apollo 14 in 1971. It consists of 100 individual reflectors, each 3.8 centimetres in diameter.

stations and the reflectors to a precision of 2 or 3 centimetres. Complex calculations which involve the gravitational pulls of the Earth, Sun and other planets yield a wealth of data about the motions of both the Earth and the Moon. Lunar laser ranging contributes about a hundred measurements a year towards UT0—the Universal Time at each of the ground stations.

Satellite laser ranging

Similar in principle to lunar laser ranging, but less demanding, is satellite laser ranging. Here distances are measured by bouncing laser pulses off artificial satellites equipped with laser reflectors. The main satellites in use are the two Laser Geodynamics Satellites, known as Lageos, launched in 1976 and 1992. The satellites are purpose-made for laser ranging, consisting of a brass core within a 60-centimetre spherical aluminium shell covered in 426 reflectors. Each has a mass of more than 400 kilograms.

They orbit at a height of 5900 kilometres, which is high enough not to be seriously affected by the remaining air in the outer reaches of the atmosphere, and are visible from all parts of the world. More than 30 ground stations track the satellites, measuring their distances to better than 2 centimetres.

The satellites orbit around the Earth's centre of mass. Meanwhile, the Earth rotates beneath them, and by careful measurements of the positions of the satellites the rotation of the Earth can be monitored. About 9000 passes are tracked each year. Satellite laser ranging is best for daily measurements of polar motion, though it is also used for short-term measurements of UT1.

GPS

The US military Global Positioning System (GPS) has revolutionised timekeeping in several ways. In Chapter 4 we saw how its "constellation" of 24 satellites is used to transfer time from the world's timekeeping centres to BIPM and we shall come back to that in the next chapter. But it is also used to monitor the rotation of the Earth. At least five satellites are visible at all times from any point in the world, and sometimes as many as eight or ten in certain locations. By processing time signals from them it is possible to determine the position of the receiver on the Earth to high accuracy. By continual monitoring from a network of 30 ground stations, IERS can track the rotation of the Earth. Using GPS, IERS measures UT1 to within 60 microseconds in the short term, and provides daily measurements of polar wobble.

DORIS

The newest technique for monitoring Earth rotation is known as DORIS: Doppler Orbitography and Radiopositioning Integrated on Satellite. DORIS relies on a worldwide network of about 50 radio beacons whose signals can be received by a suitably equipped satellite. So far only the French SPOT and US Topex satellites have DORIS equipment, but others will follow shortly. The satellite measures the frequency of the radiowaves received simultaneously from up to two beacons within sight. The waves are Doppler shifted by the motion of the satellite, and so an on-board computer can determine the speed of the satellite with respect to the beacons, and from there the precise orbit can be calculated. DORIS was originally intended to provide accurate heights for ocean survey satellites, but is now being used for other geophysical projects. Information about the rotation of the Earth is deduced in a similar way to GPS and laser ranging.

How the Earth's rotation has varied

In this section we will look at how the rotation of the Earth has varied over the years as revealed by the changing length of the day. Figure 5.3 shows the average length of the day for each year since 1700, compared with a standard day of 86 400 SI seconds, that is, exactly 24 hours. Up until the 1960s these measurements derive exclusively from astronomical observations of star transits. The most recent ones use the methods we have just described. Because these are yearly averages they do not show the faster variations which we will come to shortly. But they do reveal the medium- and long-term changes over the past three centuries. It is striking that there seems to be no pattern in these variations. If the Earth is steadily slowing due to tidal drag from the Moon, then it does not show up here. The length of the day seems to vary without any discernible pattern. The most we can say is that changes are occurring on the scale of decades which can alter the length of the day by up to 4 milliseconds.

Throughout the eighteenth and nineteenth centuries the length of the day kept close to 86 400 seconds—except towards the end of this period—and this is as expected: the ET second was defined in terms of the average mean solar second during this period and the SI (atomic) second was chosen to match the ET second as closely as possible.

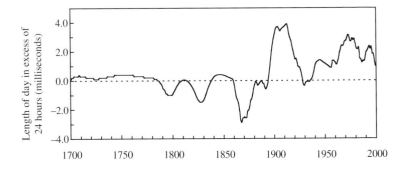

Figure 5.3. Historical records allow astronomers to reconstruct how the length of the mean solar day has changed over the past three centuries. A sudden but short-lived lengthening of the day in the late 1950s corresponds to the slowing down recorded in Figure 3.1.

Why the rotation is not constant

Long-term changes

In Chapter 1 we saw how Edmond Halley struggled to explain discrepancies between the position of the Moon in his time and many centuries earlier. It later turned out that these differences were caused by the tidal drag of the Moon on the Earth, causing the Earth to gradually lose rotational energy and speed. The world was turning faster in the past and the day was shorter.

Current best estimates are that tidal drag is lengthening the day by 2.3 milliseconds per century and has been doing so for at least 2700 years. But historical research by Richard Stephenson at Durham University and Leslie Morrison at the Royal Greenwich Observatory shows that tidal drag by itself is not enough to account for all the changes over that period. By studying timings of eclipses from as long ago as 700 BC, they found that at least two other effects seem to be at work as well.

One of these "non-tidal" effects tends to *shorten* the day by an average of 0.6 milliseconds per century. This process is a bit of a mystery, but a plausible explanation is that the shape of the Earth has been changing since the end of the last ice-age some 10 000 years ago. As the ice sheet retreated towards the poles, the newly exposed land expanded upwards as the overlying weight was removed. The effect of this "post-glacial rebound" is that the Earth is becoming more spherical. Its equatorial bulge is being pulled in, and as it does so its rotation tends to speed up.

A second effect is even more mysterious. The Earth appears to be speeding up and slowing down over a period of about 1500 years, varying the length of the day by about 4 milliseconds either way. Among the ideas put forward to explain this are movements in the core of the Earth (see the next section) and changes in sea level.

Taken together, these three effects cause the day currently to lengthen by an average of 1.7 milliseconds per century. But from the figure we see little evidence of this slowing. The reason is that the diagram is dominated by much larger fluctuations that occur over much shorter intervals. In the long run, tidal drag is slowing the Earth but as far as present-day timekeeping is concerned its effects are almost entirely negligible.

Random variations

The most obvious variations occur on scales of decades, and these are the fluctuations that cause the most trouble to timekeepers. If you look back to Figure 3.1, which shows how the length of the UT second was changing between 1955 and 1958 while NPL and USNO were calibrating the atomic second, you can see that the apparently straight line corresponds to a short and wholly untypical section of this diagram where the Earth was slowing unusually quickly. Shortly afterwards it started speeding up again before embarking on another slowing that lasted to the mid-1970s. In the 1990s the Earth was speeding up again and in the middle of 1999 the length of the day briefly touched 86 400 seconds exactly.

Data from before 1800 are rather uncertain, but what are we to make of the deep dip in the record in the mid-1800s and the even larger peak in the early 1900s?

It now seems that the origins of these fluctuations lie deep within the Earth, at the boundary between the core and its mantle (Figure 5.4). The Earth's core is made of iron and nickel, and at the temperatures and pressures prevailing at these depths the outer part of the core is molten. The liquid metal can flow, and indeed circulating flows in the core

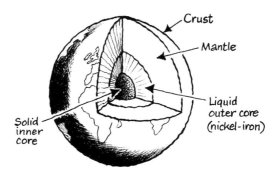

Figure 5.4. Slow and irregular changes in the length of the day are believed to be caused by movements in the liquid outer core below the Earth's mantle.

give rise to the Earth's magnetic field. By studying small changes in the magnetic field, geophysicists can work out what is happening in the core. It turns out that these motions are very slow, less than a millimetre per second, but they are irregular, and the idea is that their magnetic fields will pull and push on the overlying solid mantle, tending to speed it up or slow it down. There may even be upside-down "mountains" on the lower surface of the mantle which the molten metal pushes against as it flows past. At present there does not seem to be any way in which we can predict these flows and the consequent fluctuations in the Earth's rotation.

Seasonal variations

As we saw in Chapter 1, the seasonal fluctuations in the length of the day were discovered in the 1930s, and were really the final straw for timekeeping based on the mean solar day. Figure 5.3 contains yearly averages so these variations do not show up. But the next diagram (Figure 5.5) shows the day-to-day changes in the length of the day for the last three years. We are looking here at the very end of the curve in Figure 5.3, where the day is steadily shortening.

The shortening is apparent in this graph also, but now we see clear evidence of a regular change through the year. The size of this seasonal variation is comparable to the longer-term changes, but it is much faster.

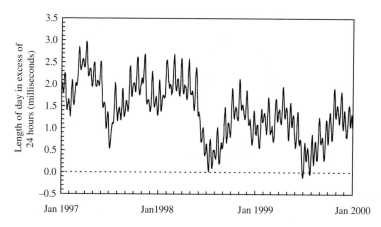

Figure 5.5. Daily measurements of the length of the day are plotted for the years 1997 to 1999. A gradual speeding up is discernible, as are the seasonal variations through the year. The rapid fluctuations are caused by lunar and solar tides distorting the shape of the Earth. The length of the day fell below 24 hours on two occasions, in June and July 1999.

The variation is double-peaked, with the Earth turning slowest in spring and autumn, and speeding up sharply in mid-summer. A lesser speeding up is apparent around New Year, but this is not so clear.

It is now certain that these variations are caused by the changing atmospheric circulation through the year. As weather patterns alter, the directions and strengths of the winds also change. Winds blowing against mountains are quite capable of changing the Earth's rotation by a measurable amount. Fluctuations on the scale of 30–60 days are also apparent and they too are thought to be due to winds. Measurements of changes in the angular momentum of the atmosphere match the changes in length of day very closely.

Ocean currents probably play a small part as well, though they are much more constant than air movements. An interesting event occurred early in 1983 when the day suddenly lengthened by about a millisecond for a few weeks. This coincided with the El Niño phenomenon, where ocean currents in the eastern Pacific reverse direction, accompanied by unusual patterns of atmospheric pressure.

Figure 5.5 also shows much faster regular variations with periods of two and four weeks. These are caused by tides raised in the Earth by the Sun and the Moon. As the tides distort the shape of the Earth, the period of rotation changes too.

The leap second

So far we have been discussing how the length of the day varies. The fluctuations seem rather small, and have never differed more than 4 milliseconds from 24 hours over the past 300 years. A millisecond is indeed a very short time. A point on the Earth's equator would turn about half a metre in a millisecond, and we might wonder whether the effects on timekeeping are equally small. In fact, they are considerable.

Let us suppose the day is a constant 1 millisecond longer than 24 hours. And let us also suppose we check the time kept by the rotating Earth—UT1—every day against an atomic clock ticking SI seconds. After 1 day UT1 will lag by 1 millisecond. The next day it will lag by 2 milliseconds. After a year it will lag by 365 milliseconds, and the difference is becoming noticeable. After 10 years the Earth clock will be lagging 3.65 seconds behind the atomic clock.

Figure 5.6 shows the cumulative difference between UT1 and TAI recorded over the past 300 years. Of course, atomic time has only existed since 1955 (and only called TAI in retrospect) but it is possible to track the difference by using Ephemeris Time for the earlier dates.

Here in one diagram is *the* problem with atomic timekeeping. A perfectly accurate clock, ticking SI seconds and set to TAI in 1958, would have been running at least 15 seconds behind UT1 during the eighteenth century and 20–40 seconds behind during the nineteenth. Yet in the twentieth century it would appear to have speeded up from 35 seconds behind in 1900 to more than 30 seconds *ahead* at the end of the 1990s. No wonder astronomers have abandoned timekeeping based on the Earth's rotation! We see also why TAI alone is not a good time scale for precision timekeeping and why it was so difficult for timekeepers to reconcile TAI with UT1 in the 1960s.

The solution, effective since 1972, is to insert a leap second into

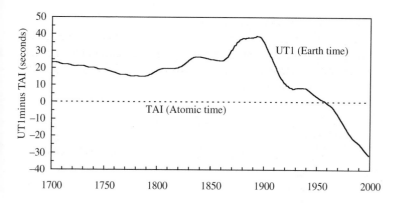

Figure 5.6. The changes in the length of the day cause UT1 (time kept by the rotating Earth) to differ from the constant rate of TAI (atomic time). Astronomers have to take these differences into account when using old observations.

UTC to keep it at all times within 0.9 seconds of UT1. The decision as to when a leap second is needed is taken by the IERS and announced some months in advance. A leap second may be either positive (making 61 seconds in the minute) or negative (59 seconds in the minute), but all 22 have been positive so far. IERS is authorised to introduce leap seconds at the end of any calendar month, but preference is always given to the end of December and June, and all leap seconds to date have been at the ends of these months. Figure 5.7 shows a leap second caught on an atomic clock at NPL at the last instant of December 1998.

Figure 5.8 shows the difference between UT1 and TAI throughout the era of atomic time. This is an enlargement of the last section of Figure 5.6, but in addition it shows the value of UTC since 1972. You can see how leap seconds have been necessary almost every year since then, forming a series of steps that remain close to UT1.

When the modern version of UTC began in 1972, BIH (as it then was) was charged with keeping UTC within 0.7 seconds of UT1. This was quite a challenge. 1972 began with a one-off adjustment of just over 0.1 seconds to set UTC exactly 10 seconds behind TAI and about 45 milliseconds ahead of UT1. By the summer, UT1 had already slipped

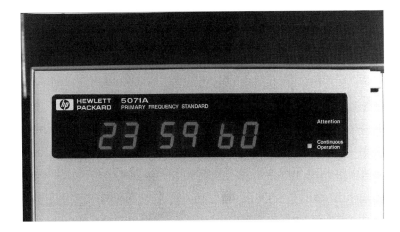

Figure 5.7. The last minute of 1998 had 61 seconds. The leap second was caught by this atomic clock at NPL.

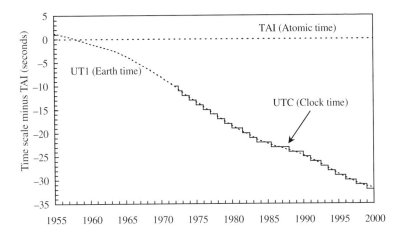

Figure 5.8. This enlargement of the last part of Figure 5.6 shows how, since 1972, UTC has been kept within 0.9 seconds of UT1 by the insertion of occasional leap seconds.

more than 0.6 seconds behind UTC and the first leap second was introduced at the end of June. At the end of the year, UT1 was starting to fall behind again and a second leap second was thought necessary, but that brought UT1 a full 0.81 seconds ahead of UTC which was technically out of bounds. The limit was relaxed from 0.7 to 0.9 seconds in 1974, and since then UTC has never departed more than 0.78 seconds from UT1, and that was in June 1994.

The IERS publishes its continuous monitoring of the Earth's rotation in its Bulletin A, a twice-weekly report that is as comprehensible to the casual reader as BIPM's Circular T. If you want to see one it can be downloaded from the IERS website (see Appendix). Each Bulletin A contains preliminary results for the previous few days; namely the position of the pole and the difference between UT1 and UTC, calculated from the methods described earlier in this chapter. At present, daily values of UT1–UTC and the length of the day can be measured to within a few microseconds. Bulletin A also contains forecasts of UT1–UTC for up to a year ahead, though the uncertainty mounts rapidly the longer the prediction. At the time of writing, in the spring of 2000, Bulletin A reveals that the divergence between UT1 and UTC has slowed, and it is clear that UTC will still be comfortably ahead of UT1 by the end of 2000 and no leap second will be needed before the beginning of 2001, if then.

Every month IERS issues two months' worth of final results in Bulletin B, which also lists measurements derived from each of the observing techniques that have been used in that period. When IERS decides a leap second will be needed, it issues the announcement some months in advance in its Bulletin C.

How many leap seconds do we really need?

It is often said that leap seconds are necessary because the rotation of the Earth is slowing. A casual reading of this chapter and other literature about leap seconds may indeed lead to the notion that the Earth is slowing down and leap seconds are needed to give it time to catch up. This is not the case. For most of the 1990s the Earth has been speeding up (from year to year anyway) and is now at its fastest for 60 years. Yet we have had 22 leap seconds in the past 27 years. So why, really, do we need leap seconds?

You may remember at the beginning of Chapter 4 we pointed out the problem of building identical clocks. No matter how carefully they are built, two clocks will always run at slightly different rates and so, in time, their readings will diverge noticeably. The principal reason that UT1 diverges from TAI, and the reason we need leap seconds, is because they have been running at different rates for several years—UT1 ticks more slowly than TAI. The mean solar second of UT1 since 1955 has been somewhat longer than the atomic SI second of TAI. This is unavoidable, since the atomic second was carefully chosen to correspond to the second of Ephemeris Time, which was approximately the average length of the mean solar second in the eighteenth and nineteenth centuries. It follows, however, that UT1 and TAI would have diverged less if the SI second had been defined differently.

Suppose the SI second, defined by the caesium transition, had been chosen to be equal to the mean solar second on 1 January 1958, when TAI was set equal to UT1. Instead of being defined as 9192 631 770 periods of the caesium transition it could have been defined as about 9192 631 937 periods, the length of the mean solar second in January 1958 (Figure 3.1). That is a tiny difference—less than two parts in 100 million—but how would our timekeeping have been affected?

Figure 5.9 shows the difference. By 1999 UT1 would have been barely 8 seconds behind TAI instead of more than 31, and since 1972 we would have needed only six leap seconds instead of 22. And that dramatic difference is caused merely by defining the second to agree with UT rather than with ET. Unfortunately that option was no longer available in 1958. By then the ET bandwagon was rolling and the CGPM was already committed to an SI second based on it.

We can do even better, with the benefit of hindsight this time, by asking what length of second would have minimised the discrepancy between TAI and UT1. That turns out to be 9192 631 997 caesium periods. With the second defined that way, UT1 would always have been within 2 seconds of TAI since 1955—sometimes ahead and sometimes behind—and there would have been only three leap seconds since 1972, two negative and one positive. And with the rates of TAI and UT1 much closer, these leaps would have been genuine consequences of the changes in the length of the day.

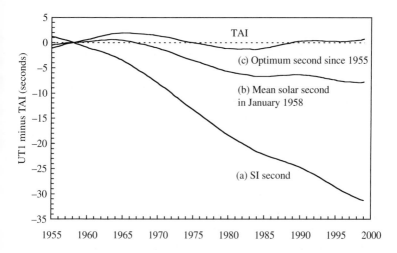

Figure 5.9. TAI was defined to be equal to UT1 in January 1958. Since then, the two scales have diverged because the SI second (a) has been slightly longer than the second of UT1. If the SI second had been set equal to the second of UT1 in 1958 (b), the divergence would have been less. With the benefit of hindsight, we can even define an optimum second (c) that would have kept UT1 within 3 seconds of TAI since 1958. The frequent occurrence of leap seconds in UTC is primarily due to the chosen length of the SI second, not changes in the rotation of the Earth.

This is not the whole story of course. The apparent convenience of maintaining TAI close to UT1 since 1955 is—alas—countered by the havoc wreaked on the history of astronomy. From Figure 5.6 we see that in 1700, for example, UT1 was 23 seconds ahead of TAI. With our redefined 1958 second that would have changed to 123 seconds *behind* TAI and with our optimum second it would have been 178 seconds behind—almost three minutes. So on balance, it may be better to leave time as it is.

6

TIME TRANSFER

Finding the time

Sometimes all this talk about the definition of time—whether on the rotation of the Earth, the length of the year, or the vibrations of caesium atoms—seems remote from daily life. If you or I want to know the time we don't observe the passage of stars with a transit telescope, we don't take photographs of the Moon over several years to work out what Ephemeris Time was when we started, and we don't take weighted averages of a couple of hundred atomic clocks around the world. We look at a clock.

We trust a clock—whether a wristwatch, a mains-powered electric clock, or an ancient grandfather clock—when we believe it reads the right time. And we believe that for two reasons. First, because we have some experience of the past behaviour of the clock—that it is a "good timekeeper". With modern quartz-crystal clocks perhaps we don't even need that experience. We take it for granted that it will not run fast or slow. But why does our clock run at the rate it does, and how do we know the rate is correct? Because at some time in the past it has been tested against a superior clock which does run at the right rate. Perhaps the manufacturer did it, as with a quartz-crystal watch, or perhaps we did it ourselves last time we checked it.

This brings us to the second reason we believe the time on the clock. Every clock has been set to read the correct time by comparison with another, more reliable clock. If the time was set correctly last week then, provided our clock is a good timekeeper, we can trust that it is still reading the correct time today.

So we see that there is a hierarchy of clocks: I set my clock by looking at your superior clock, you set your clock by an even better clock, and so on all the way back to the UTC time scale disseminated from BIPM. This chapter is about the distribution of time, how time can be transferred from place to place, and how the hierarchy has been flattened to the point where every one of us can tap into UTC only one or two steps removed from Paris.

Three principles of time transfer

There are many ways of getting access to accurate time, but they all depend on one of three basic principles. Before we look at some of the ways in which you and I can obtain time to an extraordinary degree of accuracy, we will look at these principles in some detail.

One-way time transfer

By far the most common way of transferring time from one clock to another is the one-way method. At its simplest, I set my watch by looking at the kitchen clock. I look at the clock, I look at my watch and adjust its hands until the watch and the clock read the same. I have completed a one-way time transfer from the clock to my watch. But what, say, if I wanted to set my watch by a clock out of sight in the next village? Now things get more complicated. How do I know what the distant clock is reading? And when I find out, how do I take account of the delay in transferring time from there to here?

Suppose Anne lives in a remote village where the only reliable source of time is the church clock. The clock strikes every hour and Anne and the other villagers set their own clocks by listening for the hourly "bong". This simple example of one-way transfer works well so long as no one cares whether the time is particularly accurate.

Anne realises this is not good enough. Not only is the church clock not the best of timekeepers, but villagers on the outskirts of the village hear the bong several seconds after those near to the church. She decides to invest in her own clock which keeps time much better than the church clock. The word gets round about the superior clock which has arrived in the village and the neighbours ask if they can set their own clocks by it.

Her first thought is to broadcast her own time signals. She sets up a cannon on the roof, and wires it up to her clock so that the village hears

Figure 6.1. One-way time transfer. Anne's accurate clock sets off a time gun at noon every day. Bill can set his clock by the blast, but he has to correct for the time taken for the sound to reach him.

a blast from the gun every day at noon precisely. The villagers can now set their clocks by Anne's gun rather than the church bell.

Bill, who lives on the other side of the village, is much impressed by Anne's new time service. But he realises that the improved accuracy of the clock means that he should now take account of the time delay in sound travelling from Anne's gun to his house. Bill listens for the blast and sets his clock when he hears it. But the sound of the gun has taken several seconds to travel from Anne's house to his. Knowing the distance between the houses he can estimate the travel time—sound travels about 330 metres per second—and apply a correction to his clock. For example, if Anne's house is 3 kilometres away, the travel time will be 9 seconds, and Bill will set his clock to 12:00:09 when he hears the noon blast. If he does this for a few days he will be able to compare the rates of the two clocks as well, and adjust his clock to the same rate as Anne's (Figure 6.1).

Other people can benefit from Anne's signals as well. If they know the distance from their houses to Anne's, they too can set their clocks by this method. By this means, one-way time transfer can be used to *disseminate* time to a wider public.

The weakness in this method is the need to compensate for the travel time delay between Anne and Bill. The speed of sound can vary with the

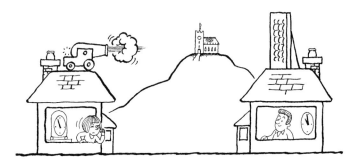

Figure 6.2. Two-way time transfer. By measuring the time for an echo to be returned from Bill's reflector, Anne can then emit her noon signal early so Bill hears the blast at exactly the right instant.

temperature and humidity of the air, and so changes from day to day. Being of a fussy disposition, Bill wants to be sure he is getting the full benefit of Anne's accurate clock. What can he do?

Two-way time transfer

Bill decides that he needs to measure the time delay every day, shortly before the time signal. He sets up a huge reflector at his house. He persuades Anne to set off a test blast some minutes before noon and listen for the echo from Bill's reflector. If she hears it 18 seconds later, then she knows that the sound took 9 seconds to travel from her house to Bill's. So as noon approaches she arranges to set off the 12:00 blast 9 seconds early. As a result, Bill will hear the gun at noon precisely and can set his clock accordingly (Figure 6.2).

Two-way time transfer potentially offers big gains in accuracy, since the delay is actually measured rather than estimated. No matter how much the local speed of sound changes with atmospheric conditions, Anne's correction will always compensate accurately for the delay.

On the other hand, the method is customised for Bill's house. If other people wanted to use the two-way method they would have to set up their own reflectors and make their own arrangements with Anne for

Figure 6.3. Common view time transfer. Both Anne and Bill time the church clock against their own clocks and exchange their measurements. Bill can now set his clock to agree with Anne's, though he must still take account of their different distances from the church. The accuracy of the church clock does not matter.

sending test signals and applying the appropriate corrections. This soon gets expensive and complicated.

Common view time transfer

Meanwhile Anne has been thinking. Two-way transfer, she reasons, can be very accurate but the reflectors and special signals make it cumbersome. Is there a simpler and cheaper way to transfer accurate time? Anne now has a brainwave—a bit of lateral thinking. She abandons the gun altogether and reverts to the church clock, but now as a means of transferring time rather than as a source of time. Every day at noon she listens for the striking of the bell and times it with her accurate clock. Bill does the same with his clock. She then sends her timings to Bill. Suppose Anne heard the bell at 11:57:24 on her clock. Bill recorded 11:58:09, so Bill knows that his clock was running 45 seconds ahead of Anne's clock. As before, they can do the test over several days to set the rate of the clock as well (Figure 6.3).

If the church clock is exactly the same distance from Anne and Bill, the sound of the bell arrives simultaneously at their houses and no correction for travel time is necessary. If the distances are not the same, then the nearer house will hear the bong first and this difference has to be taken into account. But the time delay now depends only on the difference between the distances of the clocks from the church, which will typically be much less than the distance between the clocks themselves.

Note particularly that *it doesn't matter what time the church clock struck noon*. The accuracy of the church clock does not come into it. It is used only as an intermediary to transfer time from Anne to Bill. Another interesting feature of the common view method is that the role of the church clock is entirely passive. The church itself plays no part in the process; the priest is probably not even aware that Anne and Bill are using his clock to transfer time.

On the other hand, Anne has to take the trouble to send her timings to Bill and the time transfer cannot be completed until this has happened. This method *could* be used to disseminate time to the village—Anne could post her timings on the village notice board for all to see—but common view only works when there is active cooperation and exchange of information between the source and the recipient of the time. For this reason, unlike the one-way method, common view is not much used to disseminate time to the public.

The object of the common view need not be a clock. In the mid-nineteenth century the longitude difference between Palermo Observatory in Sicily and Lecce, 480 kilometres away in the "heel" of southern Italy, was measured by groups of astronomers lying on their backs looking for meteors. Meteors occur sufficiently high in the atmosphere to be visible from widely separated sites, and by timing the meteors' appearance with clocks set to local solar time, the astronomers could work out the difference in solar time and hence the longitude between the two locations. The result was accurate to within 4 seconds of arc.

Each of these three principles of time transfer has its place. As we shall see, the cheap and simple one-way method is widely used for public dissemination of time. Two-way transfer is appropriate when the expense can be justified by the need for highest accuracy. Common view is the method of choice for high accuracy at moderate cost, and is generally used for transferring time between timing laboratories.

Time balls and time guns

One of the first systematic attempts to disseminate time came in 1833, when the Royal Observatory in Greenwich installed a "time ball" on its roof (Figure 6.4). The observatory sits on a hill in Greenwich Park overlooking a stretch of the Thames which was then bustling with ships visiting the Port of London. Shortly before 1:00 pm a bright red ball would be hoisted to the top of a mast on the roof of the building. At 1.00 pm precisely, the ball would drop and the ships on the river could set their chronometers. The Greenwich time ball is believed to have been the first regular public time signal, and its use in disseminating time from the observatory to the world's shipping led eventually to the adoption of Greenwich Mean Time as an international standard. This was an example of a one-way method, and was successful because the time taken for the light to travel the mile or so from the observatory to the river was much, much less than the uncertainty in the time the ball was dropped.

Time transfer over much larger distances became possible with the invention of the telegraph. In 1852 the Royal Observatory in Greenwich began transmitting time signals from a master clock that were used to synchronise other observatory clocks and to drop the time ball. Signals were fed to the railway companies and over the next few years Greenwich time became available to trigger time balls all over the United Kingdom. From 1861 a time gun at Edinburgh Castle would be triggered by a signal from the nearby Royal Observatory on Calton Hill. Two years later a similar gun in Newcastle was fired at 1:00 pm every day by a telegraph signal from Greenwich, 400 kilometres away. Telegraph signals also travel at close to the speed of light, so even at these larger distances the time delay was negligible.

Time by radio

With the discovery of radiowaves in the 1880s, the potential for transferring time across vast distances became apparent. In 1904 the US Naval Observatory inaugurated the world's first radio time signals from a transmitter at Navesink, New Jersey, with a regular service being broadcast from Washington DC the following year. In 1910 observatories

Figure 6.4. The time ball at the Royal Observatory in Greenwich was the world's first regular public time signal. It was dropped at exactly 1 pm every day. This multiple exposure shows the ball at four positions in its journey up and down the mast.

in France and Germany began transmitting signals, with a station on the Eiffel Tower becoming especially well known. These radio time signals, audible over continental distances, allowed astronomers to compare directly the time scales issued by different observatories. And it was the embarrassing discrepancies between them that prompted the formation of the Bureau International de l'Heure in an effort to sort out an international system for timekeeping, as we saw in Chapter 1.

Radio waves in a vacuum travel at the speed of light, some 300 000 kilometres per second. In the Earth's atmosphere they go somewhat slower—typically 280 000 kilometres per second. That means that a time signal transmitted from a suitably powerful radio station can be received on the other side of the world about 70 milliseconds later. When radio time signals were introduced, such delays were of no consequence and radio was a timekeeper's dream: accurate time could be made available instantaneously, anywhere in the world. But today, now that UTC is being realised with an accuracy of a few tens of nanoseconds, for the most demanding applications these tiny delays cannot be ignored.

Short-wave services

About a dozen short-wave radio stations broadcast continuous (or nearly continuous) UTC time signals over very large distances and are popular sources of accurate time. Their call signs and frequencies are listed in Table 6.1. Most of them transmit at or very near the standard frequencies of 2.5, 5, 10, 15, 20 or 25 megahertz. Reception is generally better on the higher frequencies during the day and on the lower frequencies at night, and best of all where both the transmitter and receiver are in daylight or darkness. In many cases the broadcast frequency is also derived from the same UTC clock as the pulses, and the stations can be regarded as accurate standards of frequency as well as time.

Perhaps the most famous of these stations is WWV, controlled by NIST in the United States. WWV began short-wave transmissions in 1923 and is now based at Fort Collins, some 75 kilometres north of the NIST Time and Frequency Division in Boulder, Colorado. Its sister station, WWVH at Kauai in Hawaii, has been broadcasting on short-wave since 1948. Both stations broadcast continuous "ticks" at 1-second intervals, with longer tones to mark the minutes and hours. The minute

Name	Location	Low frequency (kHz)	High frequency (MHz)
ATA	New Delhi, India		10
BPM	Pucheng, China		5, 10
BSF	Chung-Li, Taiwan		5, 15
CHU	Ottawa, Canada		3.330, 7.335, 14.670
DCF77	Mainflingen, Germany	77.5	
HBG	Prangins, Switzerland	75	
HLA	Taedok Science Town, Korea		5
JG2AS	Sanwa Ibaraki, Japan	40	
JJY	Sanwa Ibaraki, Japan		5, 8, 10
MSF	Rugby, UK	60	
RBU	Moscow, Russia	66.66...	
RWM	Moscow, Russia		4.996, 9.996, 14.996
TDF	Allouis, France	162	
VNG	Llandilo, Australia		2.5, 5, 8.638, 12.984
WWV	Fort Collins, USA		2.5, 5, 10, 15, 20
WWVB	Fort Collins, USA	60	
WWVH	Kauai, USA		2.5, 5, 10, 15
YVTO	Caracas, Venezuela		5

Table 6.1. Radio stations broadcasting UTC time signals. All these stations broadcast UTC continuously or almost continuously. Other stations transmit intermittently. The high-frequency transmissions can be received on normal short-wave radio sets; the low-frequency transmissions require specialised receivers and are mainly used to control automatic equipment. (*Annual Report of the BIPM Time Section 1998.*)

markers are preceded by voice announcements, either male (WWV) or female (WWVH).

The time signals are derived from local atomic clocks and transmitted with an accuracy of 0.01 milliseconds compared with UTC(NIST). Where accuracy is critical, the time taken by the pulses to travel from the transmitter to the receiver has to be taken into account. A delay of about 1 millisecond is introduced for each 280 kilometres the radio signal has to travel and this has to be corrected for if accurate UTC is required.

For example, the WWV transmitter in Fort Collins is about 7500 kilometres from NPL in Teddington. If the signals took the most direct route they would arrive at NPL about 27 milliseconds late. In practice, they travel by a varying path involving multiple reflections between the surface of the Earth and the ionosphere. The height and density of the ionosphere varies with season, time of day and even magnetic activity on the Sun, so by the time the signals arrive at NPL their emitted accuracy of 0.01 milliseconds is severely degraded. Even with the best possible estimates of the path length, the received pulses cannot give UTC to an accuracy of better than 1 millisecond. Nonetheless, this degree of accuracy is more than enough for most purposes.

The common view method can be applied to radio signals, just like the church clock. In Chapter 3 we saw how two teams of scientists, on opposite sides of the Atlantic, worked together to calibrate the atomic second in terms of Ephemeris Time. William Markowitz's group at the US Naval Observatory was using the Moon camera to measure the length of the UT2 second in terms of ET. They were faced with the problem of how to transfer UT2, determined at USNO and kept by the observatory's quartz-crystal clocks, across the Atlantic to Louis Essen's group at NPL where it could be compared to a similar crystal clock acting as the "face" of the caesium standard. The solution was to use a common view method using WWV.

Once a month, both groups tuned in to the 1-second pulses from WWV, then based in Greenbelt on the outskirts of Washington DC. USNO timed the pulses in terms of UT2 and NPL simultaneously timed them with respect to the caesium standard. By exchanging their timings, in a similar way to Anne and Bill, the two groups could measure the length of the WWV second both in terms of the UT2 second and in terms of the number of caesium cycles. It was then a simple matter to calculate the number of caesium cycles in a second of UT2. In that case, too, the actual length of the second transmitted by WWV did not matter, it was just used to transfer time intervals of UT2 from USNO to NPL.

An interesting point here is that WWV was very much closer to USNO than to NPL. If it had been used to transfer time rather than time interval, then the distance of USNO and NPL from the transmitter would have to be taken into account, with all the problems we have just mentioned. But because only intervals were being transferred, propagation delays did not matter so long as they remained reasonably constant.

Long-wave services

Anyone with a normal short-wave radio can pick up the pulses from WWV and the many similar time services. But towards the other end of the radio spectrum, at wavelengths too long for normal radio receivers, another set of transmissions can be found. These stations, rarely heard by the human ear, are designed to supply accurate time signals to automatic equipment.

The most familiar recipient of these signals is the increasingly popular radio-controlled clock. It is essentially a quartz-crystal clock connected to a radio receiver. Most radio clocks in the UK are tuned to the regular pulses from a station known as MSF, which broadcasts time signals derived from UTC(NPL). Along with second "ticks", MSF transmits the date and time of day in a coded form that lets the clock set itself to the correct time. Radio clocks controlled by MSF can adjust automatically to the beginning and end of summer time, and can even cope with leap seconds, though not in the most elegant fashion.

MSF is based just outside Rugby, in the English Midlands, amid a forest of radio masts between the M1 motorway and the London–Manchester railway (Figure 6.5). It is controlled by the atomic clocks at NPL some 120 kilometres to the south. (As a matter of interest, time is transferred from NPL to the MSF site by a common view method using an intervening television transmitter as the intermediary.)

MSF broadcasts on a frequency of 60 kilohertz, which corresponds to a wavelength of 5 kilometres. These very long waves, too long to be received on ordinary domestic radio sets, travel close to the ground and can follow the curve of the Earth for thousands of kilometres. This means that the signal takes a more-or-less direct path to the receiver and, unlike short-wave radio, is little affected by variable propagation conditions high in the atmosphere. Usable signals from MSF can be received over a radius of 2000 kilometres, which includes much of western Europe.

Similar low-frequency stations in Europe are HBG in Switzerland

Figure 6.5. One of the two 250-metre masts supporting the MSF transmitting aerial near Rugby. MSF transmits continuous time signals on 60 kilohertz which are widely used to control radio clocks and other equipment.

and DCF77 in Germany. DCF77 broadcasts on 77.5 kilohertz from Mainflingen, 25 kilometres south-east of Frankfurt, and is controlled by the awesome atomic clocks at PTB in Braunschweig. The signals are used throughout Europe to control railway clocks, domestic clocks, wristwatches and even traffic lights. In the US, WWVB broadcasts on 60 kilohertz from the same site as WWV and covers the whole of North America.

Because of the more stable propagation conditions at these low frequencies, time signals from these stations can be received with an accuracy of 0.1 milliseconds—ten times better than short-wave radio—once the travel time has been taken into account. Low-frequency radio transmissions are now the most important and economical means of disseminating UTC from the national timing centres.

Time by telephone

If you haven't got a short-wave radio you can call up WWV by telephone, but because of the unpredictable and complex path taken through the telephone network, the accuracy is no better than 30 milliseconds within the continental United States and much worse overseas. Nonetheless the telephone has proved to be an effective and popular means of disseminating accurate time. Two kinds of telephone time service are now in routine use.

The Speaking Clock

Most telephone companies provide some kind of "speaking clock" which you can call to find the time. In the UK, the Speaking Clock was brought in by the Post Office in 1936 because people kept bothering the operators by phoning to ask the time. A system of rotating glass disks assembled segments of a recorded voice to announce the time "at the third stroke" every 10 seconds. The glass disks gave way to magnetic recordings in 1963 and the present digital system came into operation in 1984. Now privatised and marketed by BT as "Timeline", the Speaking Clock continues to provide time checks six times a minute, interspersed with messages from a commercial sponsor.

The service is provided by two pairs of quartz-crystal clocks, one pair in London and the other in Liverpool. One of each pair is the source of the signals and the other is on standby. At 09:55 each morning the clocks are automatically compared to UTC(NPL) by signals sent by landline from the MSF transmitter and corrections made as required. BT guarantees that the time signals as received will be always within 50 milliseconds of UTC(NPL), but in practice they are expected to be within 5 milliseconds. About a million calls a week are made to the

Speaking Clock, with a noticeable increase immediately after clocks go forward or back to implement summer time.

Computer time

The wonders of the World Wide Web continue to grow. It is now possible, for example, to point your browser to the US Naval Observatory's master clock and see UTC(USNO) displayed on your computer screen. It is a very satisfying thing to do. But when was that time? It takes several seconds, at least, for a web page to be wafted across the Atlantic, so the time on the screen is not "now" but some moments ago. If we do not know what that delay is, then the fabulous accuracy of UTC(USNO) is not accessible to us, no matter how nice it looks.

This is frustrating, since the average personal computer is sorely in need of accurate time. Considering the mind-boggling processing power of even the cheapest PC, many people are surprised and disappointed by the poor timekeeping ability of their computer's internal clock. The clocks used in PCs can drift by several seconds a day and have to be reset frequently. It would be very useful to be able to connect up to an atomic clock and put them right.

Several timing centres now provide a service that does just that. NPL, for example, has a service called Truetime, designed to supply accurate time to computers. You first need to install the software, which can be downloaded from NPL's website, and then dial the Truetime phone number. The clock in the PC receives a time signal from the NPL UTC clock and is then set to within 20 milliseconds of UTC(NPL). If this accuracy is not sufficient, Truetime can work in a two-way mode. Your PC talks to the computer at the other end and between them they bounce signals back and forth to determine the time delay. The NPL computer then sends a corrected time signal that resets the internal clock on the PC to within one millisecond of UTC.

Similar services are offered by NIST in the US and PTB in Germany, as well as several other centres. Of course, once your PC clock is set accurately it will immediately drift away again, and there is little you can do about it except complain to the PC manufacturer or buy a new clock card.

Time from the sky

Time transfer by radio is fine provided you only need millisecond accuracy. If you just want the trains to run on time, controlling your railway clocks by a radio service such as MSF or DCF77 will achieve your aims beautifully and economically. But what if you need time more accurately?

The use of orbiting satellites to disseminate time has long been a goal of timekeepers. Since 1974 NIST time signals have been made available via two meteorological satellites positioned over Central America and the Pacific Ocean. The GOES satellites are geostationary, that is, they are at such a height that they complete one orbit every 24 hours and so appear almost unmoving in the sky. A ground station at Wallops Island, Virginia, transmits time signals to them which are rebroadcast to North and South America and most of the Pacific and Atlantic Oceans. Because of the distance of the satellites, the signals are advanced by 260 milliseconds on UTC(NIST) on transmission, which means they arrive back on Earth at about the right time. Specially designed receivers can provide signals accurate to 0.1 milliseconds.

But the most significant development in time transfer since the 1980s has been the growth of two satellite navigation systems, GPS and GLONASS. GPS is a network of 24 Navstar satellites operated by the US Department of Defense. GLONASS (Global Navigation Satellite System) is a similar array of satellites operated by the Russian military. With suitable receivers, a user of GPS or GLONASS can determine a position on the ground to within a few metres. Although the prime purpose of these systems is military—it was GPS that allowed UN coalition forces to advance swiftly across the wastes of the Iraqi desert in the 1990 Gulf War—the service is available for a rapidly growing range of civilian applications.

Of the two networks, GPS has been the more popular and GPS receivers are now available for many civilian purposes. GPS works by broadcasting continuous time and position signals from each of the spacecraft. They orbit 20 000 kilometres above the ground and at least five satellites are always visible from any point on Earth. By comparing the times at which the signals are received from each satellite, a GPS receiver can automatically calculate its position on the Earth to high accuracy.

GPS can work because each satellite carries an atomic clock. The first generation of GPS satellites carried rubidium clocks, but the current ones carry caesium clocks similar in design to the industrial standards discussed in Chapter 4. The clocks are continuously monitored from the GPS control centre, and corrections applied to ensure that all 24 satellites are keeping close time.

The time scale actually broadcast from the satellites is known as GPS time, and is derived from UTC(USNO). Because GPS cannot handle leap seconds, GPS time is essentially a continuation of UTC in 1980, when the system became operational. So GPS time is permanently 19 seconds behind TAI, and, in 2000, 13 seconds ahead of UTC. Disregarding the leap seconds, GPS time is required to stay within 1 microsecond of UTC(USNO), and in practice the agreement is very much better, with daily departures not exceeding 30 nanoseconds over the course of a year. Until recently, not many people had access to this accuracy due to the policy of the US Government to deliberately degrade the quality of the GPS signals available to civilians to preserve the most precise positioning for military applications. This degradation was unexpectedly removed in May 2000, much to the delight of the world's timekeepers.

The Russian system, GLONASS, is taking longer to gain acceptance and suffers from reliability problems. Like GPS, the constellation is supposed to have 24 operational satellites, but many have failed and by the spring of 2000 only ten were working. The time broadcast from the satellites' clocks—GLONASS time—is derived from UTC(SU), the version of UTC maintained in Russia. Unlike GPS, GLONASS time includes the UTC leap seconds and there has been no no intentional degradation of the signals for military purposes.

For ultimate accuracy, daily corrections to adjust GPS and GLONASS time to UTC are published in BIPM's Circular T. In the autumn of 1999, GPS time was within 30 nanoseconds of UTC (less the 13-second offset) while GLONASS time was 300–400 nanoseconds behind UTC. It appears that the uncertainty in GPS time from day to day is around 10 nanoseconds but several hundred nanoseconds for GLONASS time, a consequence of the difficulty in characterising time delays in the GLONASS multi-frequency system.

But even without the Circular T corrections, the time available from these satellite systems is far more accurate than anything from other sources. It is clear that GPS, and to a lesser extent GLONASS, are the best available means of disseminating UTC anywhere in the world.

GPS is even more useful in common view mode. Since 1995 BIPM has relied on common view tracking of GPS satellites to transfer time from the national centres to Paris. Pairs of centres track a satellite at the same time, comparing its signals to their own UTC clocks. Every six months BIPM publishes a daily tracking schedule, setting out the precise times at which satellites should be observed from each centre. Each track lasts 13 minutes and the start and finish times of the common observations are synchronised to within 1 second. Centres in Europe observe about 30 tracks each day (Figure 6.6). The observations are sent to BIPM who deduce the difference in UTC between pairs of centres and use it as part of the procedure for calculating TAI, as described in Chapter 4.

The BIPM time-transfer network is focused on three "stars" centred on Europe, North America and Asia, with the Paris Observatory (OP), NIST and the Communications Research Laboratory (CRL) in Tokyo at the centre of each. The main links, between OP and NIST and between OP and CRL, are monitored routinely for ionospheric delays. Experience has shown that the degradation formerly imposed on GPS signals for military purposes could be almost entirely circumvented when strict common view is used. Time transfers from a single common view observation can be accurate to 3 nanoseconds for continental distances and 5 nanoseconds for intercontinental distances.

BIPM also publishes a tracking schedule for GLONASS and collects common view data from about ten centres, but these are not yet used in the formation of TAI.

Errors in the position of the ground antennas are the biggest source of systematic error in GPS or GLONASS timings. Positions need to be known to about 30 centimetres if 1-nanosecond accuracies are being aimed at. The positions of all the timing centres contributing to TAI have been determined by comparison with the IERS Terrestrial Reference Frame which we met in the last chapter.

Figure 6.6. An array of GPS receivers being tested on the roof of the time-transfer laboratory at NPL. The dish at the right is for two-way time transfer via geostationary communications satellites.

Time transfer and relativity

Such is the accuracy with which time can be generated and transferred from place to place, that modern timekeepers have to contend with complications that simply did not arise a generation ago. Foremost of these are the subtle but real effects predicted by Albert Einstein's two theories of relativity.

Relativity has many surprising and far-reaching things to say about the nature of time and space, but for timekeeping purposes the special theory deals with clocks that are moving with respect to each other and on a rotating planet, and the general theory deals with clocks operating in a gravitational field. In both cases it is no longer possible to ignore these small effects when transferring time from one place to another, for example when BIPM computes TAI from clocks around the world.

Let's look at this in more detail. There are three relativistic effects we need to consider. The most famous of these is time dilation, summed up in the phrase, "moving clocks run slow". A clock in an aircraft, for example, would be seen to run slow as judged by an observer on the ground. (Equally, clocks on the ground would appear to run slow as seen from the aircraft, but we do not have space in this book to discuss the subtleties of relativity!) Time dilation only becomes appreciable at speeds close to that of light—indeed at light speed time stops altogether—but with the nanosecond accuracy now possible with modern atomic clocks, time dilation has to be taken into account whenever clocks are moved. Since all the primary standards (as well as the 260 secondary clocks) remain fixed, none of their readings need to be adjusted for time dilation when their readings are combined to form TAI. But there is a second relativistic effect which certainly does need to be taken into account.

Einstein showed that a clock in a gravitational field will appear to run slow compared to a clock which is in free space. A clock at the bottom of a valley, for example, will run more slowly than an identical clock on top of a mountain, because the former is closer to the centre of the Earth and so in a stronger gravitational field. This effect, known as the gravitational shift, is nothing to do with the mechanism of the clock itself, but is a characteristic of time and space.

The importance of this effect was anticipated in 1980 when a CIPM working group recommended that the scale interval of TAI should be understood as being the SI second "as realized on the rotating geoid". The "geoid", in geophysical language, is the approximate equivalent of mean sea level. If the Earth were made of water then its surface would be the geoid. The geoid is not spherical because the Earth is slightly flattened at the poles, measuring 42 kilometres less from top to bottom than across the middle, so to speak. It is also affected by the uneven

Figure 6.7. High clocks run fast. Two clocks adjusted to tick at exactly the same rate will diverge when one is taken to the top of a mountain. The observer in the valley will judge the mountain clock to be running fast while the observer on the mountain will judge the valley clock to be running slow. The difference has to be taken into account when combining measurements from atomic clocks around the world.

density of the Earth's crust and even local features such as mountains. It is specified as "rotating" because the rotation of the Earth modifies the geoid from where it would be if the Earth were not spinning. The essential point is that the Earth's gravitational "potential"—a measure of how the gravitational pull varies with distance—is the same at all points on the geoid. That means that two identical clocks placed anywhere at rest at sea level would tick at the same rate, relativistically speaking. The practical consequence for timekeeping is that data from the primary standards have to be corrected for height above sea level before they can be used to form TAI. Such is the accuracy of modern primary standards that this correction is now appreciable (Figure 6.7).

To a good approximation, a clock will appear to speed up by a factor of about 1.09 parts in 10^{13} for every kilometre above sea level, equivalent to 9.43 nanoseconds per day (at heights greater than 25 kilometres a more complicated formula is needed). NIST-7, the US primary standard we met in Chapter 4, is located at Boulder, Colorado, some 1650 metres above mean sea level. Compared to a clock at sea level, the length of the second produced by NIST-7 is therefore shorter than the SI second by a factor of 1.80 parts in 10^{13}, amounting to 15.6 nanoseconds per day. This tiny difference has to be taken into account at BIPM when EAL is calibrated to form TAI. While seeming minuscule, the correction amounts to some 45 times the present uncertainty in the TAI second.

Clocks on board orbiting satellites are affected by both time dilation and the gravitational shift, on account of their speed and their height above sea level. The orbital speed of GPS satellites ensures that an observer at sea level will see them run slow by 7.1 microseconds a day. On the other hand, their height above the Earth's surface—20 000 kilometres—causes them to run fast by 45.7 microseconds a day. The two effects together cause GPS clocks to run fast by 38.6 microseconds a day. Their designers anticipated this and set the clocks to run slow by the same amount, so that the time interval received from GPS at sea level is very close to the SI second. Small changes in speed and height caused by the orbits not being precisely circular are corrected by the GPS receiving equipment. Interestingly, GLONASS orbits are more closely circular than GPS and these additional corrections do not have to be programmed into GLONASS receivers.

A third consequence of relativity, which is rather odd and not as widely known as the other two, is called the Sagnac effect. Suppose we have two ideal clocks, set to tick at the same rate, and place them on the equator at sea level. Now we take one of the clocks on a slow sea journey eastward around the equator, being careful never to depart from the geoid. The journey has to be slow to reduce any time dilation to a negligible level and could take many years, if we so wished. When the travelling clock met the stationary clock again, after it had made a complete circuit of the equator, how would their times compare? If the journey had indeed been slow enough, and the clocks had stayed at sea level, we would expect them to read the same, but they would not. The travelling clock will be 207 nanoseconds behind the stationary clock. Even stranger, suppose the travelling clock went the other way, completing a circuit of the equator in a westerly direction. Now the travelling clock will be *ahead* of the stationary clock, again by 207 nanoseconds.

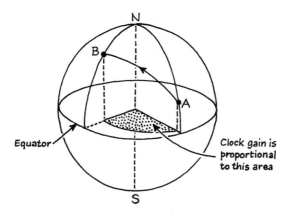

Figure 6.8. The Sagnac effect. A clock moved from point A to point B on the surface of the Earth will accumulate a time difference compared with a stationary clock in proportion to the shaded area. The area is counted as positive (a gain) for westward journeys and negative (a loss) for eastward journeys. The gain amounts to 1.62 nanoseconds for every million square kilometres.

It turns out that any journey on or near the Earth will accumulate a Sagnac timing error proportional to the area swept out by the journey projected on to the plane of the equator (see Figure 6.8). The Sagnac effect can be avoided by transporting clocks only along lines of longitude, which would normally mean going via the north or south poles. Although it appears mysterious, the Sagnac effect is really time dilation in disguise—"stationary" objects on the surface of the Earth are not stationary at all, but are rotating at a speed which depends on their latitude. We should strictly compare moving clocks with a stationary clock at the centre of the Earth, but as this is impractical we treat clocks on the ground as stationary and the Sagnac correction then makes sure the calculations come out right.

As an example of how relativity can affect practical timekeeping, let us suppose the staff at PTB in Braunschweig want to send a clock to USNO in Washington DC to compare it with the clocks there. They decide to fly it from Berlin to Washington, a distance of 6700 kilometres.

For the sake of simplicity we assume it can be done in one hop at a constant speed of 800 kilometres per hour and a constant height of 10 kilometres. On arrival at Washington the PTB clock would have lost 8.3 nanoseconds due to time dilation, gained 32.9 nanoseconds due to the gravitational redshift and gained 18.8 nanoseconds due to the Sagnac effect. If the clock was reading UTC when it started, it would read 43.4 nanoseconds ahead of UTC when it arrived. If it were then reset to UTC and flown back to Berlin, all else being equal, the Sagnac effect would act in the opposite sense and by the time the clock arrived in Berlin it would be only 5.8 nanoseconds ahead of UTC. So it's true that jetlag is worse flying westwards ....

The six pips

To conclude this chapter we will look at an example of how time dissemination has evolved over more than seven decades. Ever since 1924, BBC radio has broadcast a time signal commonly known as the "six pips". The pips originally came from the Royal Greenwich Observatory (RGO) and are known as the Greenwich time signal.

When the signals began in 1924, they were controlled by a Shortt free pendulum clock, similar to that described in Chapter 2. The clock kept GMT, or what we would now call UT0, determined from observations made with the observatory's transit circle (Figure 5.1). The "six pips" heard on the radio came from the Shortt's slave clock and were formed by an electrical contact closed by a specially made mechanism. The first five pips counted down the seconds to the sixth pip, which marked the time signal.

This arrangement lasted until 1949, when the Shortt was replaced with a quartz-crystal clock at RGO's new home at Herstmonceux Castle in Sussex. The one-kilohertz frequency from the clock was used to drive a "phonic motor" which, through a series of gears, turned a disk spinning once a second. A beam of light reflected from a silver patch on the disk created flashes of light at one second intervals which were detected electronically and used to create the pips.

It was not until 1967 that RGO acquired a caesium clock—a Hewlett-Packard 5060A—though the rotating disk was still the source of the pips. From 1972 the pips marked UTC rather than GMT, and listeners to the BBC at midnight became used to hearing a seventh pip whenever a leap second was introduced. At the same time the final pip was lengthened to remove any doubt about which pip was the signal. The caesium clock gave way to a GPS receiver in the 1980s, but the system of phonic motor and rotating disk continued in service until 1990, when RGO was relocated to Cambridge and took no further part in public timekeeping.

Responsibility for the six pips then passed to the BBC. The source of today's time signals is two GPS receivers—used to generate 1-second UTC pulses—and an MSF receiver—used to label them with time-of-day information. Radio broadcasts follow a complex path from studio to the many transmitters and the timing of the pips is adjusted so that they will be emitted as near as practical to the correct time. If you want the best accuracy from the signals, they are optimised for the 198 kilohertz transmitter at Droitwich which broadcasts BBC Radio 4 on long wave. In the UK, the signals can be received with an accuracy of better than 50 milliseconds.

So not only do the "Greenwich" time signals no longer mark GMT—as we found in Chapter 4—they lost their tenuous links with Greenwich many years ago. And as a further disappointment to those who believe the British to have a proprietary claim over the source of time, the signals broadcast from the BBC do not even originate in Britain: they are derived from GPS satellites which are in turn controlled by the atomic clocks of the US Naval Observatory in Washington DC.

7

USES OF ACCURATE TIME

Time for everyone

What is the use of accurate time? Why should anyone need to know the time to millionths or even billionths of a second? On the face of it, this degree of accuracy seems extravagant and quite irrelevant to everyday life. Surely only a few experimental scientists need to measure time to such precision? Yet, as we shall see in this chapter, the economy of the developed world is coming to rely ever more heavily on accurate time generated by atomic clocks. Without those clocks, much of modern life would simply not be possible.

Consider this. Thanks to atomic clocks, backed up by quartz-crystal oscillators, time and frequency can be measured more accurately and more cheaply than any other physical quantity. Time can be measured a thousand times more accurately than distance and millions of times more accurately than mass and, for the same levels of accuracy, at less than 1 percent of the cost of measuring either. If you want to measure something accurately, try turning it into a time or a frequency.

Accurate time is no longer hard to come by. Until recently, if you wanted really accurate time, you had to invest in your own atomic clock, perhaps one of the industrial standards discussed in Chapter 4. Today, many of the applications that once required local atomic clocks can be done more conveniently using terrestrial radio signals or GPS, with the curious consequence that extremely accurate time is more readily available than time of moderate accuracy. We have already seen how the BBC radio time signals are now derived from GPS and that trend is continuing. Radio and GPS can be so cheap and convenient that they are being used not always for their great accuracy but for their availability. Why set your own clock when a radio transmitter or satellite can do it for you?

A consequence of this is that accurate time is now available to everyone, anywhere in the world. Time signals wash around the planet 24 hours a day. Even the keepers of time, the scientists who run the national laboratories, cannot know who uses their time signals and for what purpose. So in this chapter we can do little more than touch on a few examples of the uses of accurate time, knowing that there will be countless others that we have overlooked.

The nature of time and space

The first application of accurate clocks is the investigation of time itself. We indicated in the last chapter that atomic clocks are now so accurate that tiny corrections for the effects of Einstein's theories of relativity have to be taken into account as a matter of routine. For the same reason, atomic clocks can be used as tools to investigate and test the predicted consequences of relativity and so learn more about the nature of time and space.

One of the weirdest predictions of the special theory of relativity is time dilation. A moving clock will run slow in comparison with a stationary clock. In other words, if two identical clocks are synchronised and one of them is then flown around the world and brought back to its starting point, the travelling clock will have lost time on the stationary one because it has been moving. Although the phenomenon of time dilation was predicted by Einstein in 1905, it was not until the 1960s that atomic clocks became portable and accurate enough to actually observe the phenomenon. Plenty of people had written about the discrepancies between clocks moving at different speeds, but no one had ever done the experiment. No one had ever compared the times on real clocks. Time dilation was certainly accepted by physicists—numerous experiments with cosmic rays and particle accelerators had proven it many times over—but Einstein had talked about clocks, and more than six decades later it was high time that someone actually demonstrated it with clocks.

The gravitational shift, too, cried out for a clock test. Einstein had predicted it in 1907, but it took until 1960 for physicists to show experimentally that time does indeed run slower in a gravitational field. In a particularly elegant experiment, two physicists at Harvard University—

Robert Pound and Glen Rebka—measured the tiny rise in the frequency of gamma rays emitted from a 22.5-metre high tower and detected at the ground, and the equally tiny drop in the opposite direction. An improved version of the experiment in 1965 agreed with predictions to within 1 percent—and has not been bettered—but that was not the same as seeing a high clock run fast and a low clock run slow.

Special relativity predicts that "moving clocks run slow", but would a real clock really run slow? General relativity predicts that "high clocks run fast" but would a clock on a mountain really run faster than a clock in a valley?

In October 1971 scientists from the University of St Louis and the US Naval Observatory performed the first convincing demonstration with atomic clocks of both predictions of relativity. They took a set of four caesium beam clocks around the world on commercial aircraft: first eastwards (a total of 41 hours flying time) and then westwards (49 hours). According to relativity, the clocks would have been affected by time dilation due to the speed of the aircraft (losing time), a gravitational shift due to the height of the aircraft (gaining time), and a Sagnac effect according to whether the clocks flew eastwards (losing time) or westwards (gaining time). Relativity predicted that there would be a net loss of 40 nanoseconds on the eastward trip and a net gain of 275 nanoseconds on the westward trip. The results—a loss of 59 nanoseconds and a gain of 273 nanoseconds—agreed with predictions to within the experimental uncertainty of around 20 nanoseconds. The Hafele–Keating experiment, as it was called after the two principal scientists, has since become a classic of relativity.

Nonetheless, it could not compare with the Pound–Rebka experiment for testing predictions of the gravitational shift and four years later scientists from the University of Maryland devised a more sophisticated experiment, this time with a focus on the gravitational shift. They prepared two identical packages of atomic clocks, each containing three HP 5061A caesium beam clocks and three rubidium clocks. One package was installed on a US Navy aircraft while the other stayed on the ground. On five occasions between September 1975 and January 1976 the aircraft flew for 15 hours averaging a speed of 240 knots and a height of 30 000 feet. It flew a continuous figure-of-eight path, so eliminating the Sagnac

effect, and the low speed kept time dilation to a minimum. Throughout
the flight the position and speed of the aircraft was monitored by ground
radar, while timing information was exchanged by receiving and reflect-
ing pulses from a laser on the ground. During a typical flight the airborne
clocks should have gained 52.8 nanoseconds due to the gravitational shift
and lost 5.7 nanoseconds because of time dilation, a net gain of 47.1
nanoseconds. The measured change was a gain of 47 nanoseconds with
an uncertainty of 1.5 nanoseconds, convincing evidence that high clocks
run faster than low clocks and by the predicted amount.

But the most spectacular confirmation of the gravitational shift
came in June 1976, when a NASA spacecraft called Gravity Probe
A was launched on a rocket to a height of 10 000 kilometres before
falling back to Earth.   The probe carried a hydrogen maser clock
constructed by physicists Robert Vessot and Martin Levine at the
Smithsonian Astrophysical Observatory.   As usual, two relativistic
effects were operating: time dilation due to the speed of the rocket and
the gravitational shift due to the height above sea level. By monitoring
the speed of the rocket throughout the 2-hour flight, the physicists were
able to separate out the two effects and show that at the maximum height
the gravitational shift was causing the clock to run fast by four parts in
10^{10}, as predicted by general relativity. The agreement was within 70
parts in a million.

And to clinch the claim about clocks in valleys and on mountains,
in 1977 Japanese scientists carried an HP 5061A on two return trips between astronomical observatories at Mitaka and Norikura, which differ
in height by 2818 metres. The clock spent a week in each location and
was compared with an identical clock which remained at Mitaka. Again,
the high clock was found to have gained on the low clock by the amount
of the gravitational shift predicted by relativity.

Another prediction of general relativity, more telling than the gravitational shift, is that massive objects like the Sun distort the space around
them. As a result, a beam of light from a star passing close to the
Sun will be bent and the star will appear to be displaced from its usual
position. The bending of light from distant stars was first observed by
the astronomer Sir Arthur Eddington during a total eclipse of the Sun
in 1919, and has been confirmed several times since, notably with radio

Figure 7.1. John Davis, of NPL, prepares an atomic clock for a flight between London and Washington in a demonstration of Einstein's relativity for the BBC Horizon television programme.

sources rather than stars. But a second consequence of the bending of light, which had not been predicted by Einstein himself, was that light would be delayed as it passed the Sun. That insight had to wait until 1961, when a young physicist called Irwin Shapiro calculated that a ray of light just grazing the edge of the Sun would be delayed by 125 microseconds compared with a ray traversing an identical distance in free space.

In the 1960s and early 1970s scientists had begun to measure the delay by bouncing radiowaves from Venus and Mercury when they were on the far side of the Sun, and also by exchanging signals with interplanetary spacecraft. Results were within a few percent of predictions. But the best test came in 1976 when the US landed two Viking spacecraft on Mars. They sampled the atmosphere and the soil and transmitted many pictures and other data to Earth. In November of that year the orbit of Mars took the planet around the far side of the Sun, and by measuring the time taken for radio signals from Earth to be returned from the Vikings, Shapiro and his colleagues were able to show that the additional round-trip delay—about 250 microseconds—corresponded within one part in 1000 with the predictions of general relativity. It was the most convincing demonstration to date of the gravitational time delay.

A new series of tests of relativity became possible after 1967, when radio astronomers at Cambridge discovered an extraordinary object emitting pulses at intervals of precisely 1.337 seconds. This was the first of many "pulsars", which proved to be compact stars spinning at high speed (Figure 7.2). A pulsar is a neutron star, the remains of a star that has burned up its fuel and collapsed. It consists mainly of neutrons, is extremely dense and packs the mass of the Sun into a ball only 20–30 kilometres in diameter. The fastest pulsars spin at hundreds of rotations a second while the slowest take several seconds to complete one rotation (compare that with the Sun, which rotates once every 25 days). They have strong magnetic fields and emit beams of radiowaves from their magnetic poles. If the beams sweep over the Earth we see pulses much like the flashes from a lighthouse. So regular are the flashes, that pulsars can be used as reliable astrophysical clocks.

Several observatories around the world now make routine timing measurements of pulsars. The arrival times of pulses are recorded with atomic clocks to accuracies of a few microseconds, and these times are the raw data for numerous astrophysical investigations. By averaging over many years—and billions of pulses—the rotation periods can be determined to a few parts in 10^{14}. We shall briefly look at the time-keeping consequences of this fabulous accuracy in the next chapter.

One consequence of this high accuracy is that pulsar observations are very sensitive to Doppler shifts. The speed of the Earth around

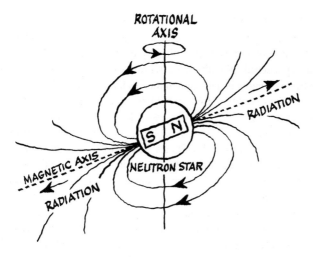

Figure 7.2. Pulsars are spinning neutron stars—the collapsed remains of burned-out stars. They emit beams of radiowaves from above their magnetic poles.

the Sun, about 30 kilometres per second, causes slight changes in the apparent frequency of the pulses which show up very clearly in the timing data. When the Earth is moving towards the pulsar the pulses come more frequently than when the Earth is moving in the opposite direction. Pulsar astronomers correct their observations for the Earth's motion as a matter of routine.

In the summer of 1974 US astronomers found that a pulsar called PSR 1913+16, ticking 17 times a second, still showed strange behaviour even after these corrections had been made. The Doppler shift was caused not only by the movement of the Earth but also by the motion of the pulsar itself. It turned out that the pulsar was in orbit around a second neutron star. The two stars swing around each other in highly elliptical orbits in a little under eight hours, a remarkably short period considering that the Earth takes a full year to make one circuit of the Sun. By carefully timing the pulses over many orbits the astronomers could map out the orbit with great precision (Figure 7.3).

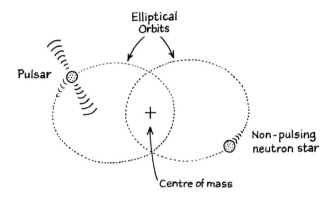

Figure 7.3. The binary pulsar PSR 1913+16 consists of two neutron stars orbiting around each other. Precise timing of the pulses allows astronomers to test Einstein's theories of relativity.

And what an orbit! The two stars are so close together that at their closest there would not be enough room to fit the Sun between them. Under these conditions we would expect relativistic effects to be prominent, and indeed they are. The orbiting clock of the pulsar shows the expected effects of time dilation and the gravitational shift, as well as another prediction of general relativity: instead of tracing perfect ellipses, the orbit does not quite close, so the path is a continuous looping pattern instead of a simple ellipse. The orbit is slewing around at the rate predicted by general relativity.

But most dramatic of all, the orbit is shrinking. One of the more exotic predictions of general relativity is that certain objects should radiate gravitational waves, ripples in the fabric of space itself. A system like the binary pulsar should lose energy by gravitational radiation and the orbit should slowly shrink. As the stars spiral in towards each other they swing around faster and faster. Relativity predicts that the period for one revolution should decrease by 75.8 microseconds each year. By the early 1990s observations showed the period decreasing by 76.0 microseconds a year with an uncertainty of 0.3 microseconds. In other words, the two stars are getting closer together at the precise rate expected from the loss of energy to gravitational waves. In fact, the two stars will collide and merge into a single object in about 300 million years.

No one has ever seen a gravitational wave, though many groups around the world have been looking for them. Yet so compelling is the evidence from the binary pulsar that its discoverers, Russell Hulse and Joe Taylor, were awarded the Nobel Prize for Physics in 1993. And all this was achieved by accurate timing of pulses from the sky.

101 uses for an atomic clock

As far as more general uses of accurate time is concerned, we can divide them into those where time or frequency needs to be known to high precision, and the special class of uses of navigational satellites to measure accurate positions on the Earth. We will return to the latter shortly, but in this section we will first look at some direct applications of accurate time.

Precise frequencies

The simplest class of uses is not to provide time at all, but simply a frequency of great accuracy. For example, manufacturers of electronic equipment must have access to a source of accurate and stable frequency, such as a caesium clock, to use as a standard of comparison when calibrating their products. Several of the radio stations which transmit time signals also have tightly controlled broadcast frequencies and can themselves be used as frequency standards. One of the most stable transmitters in the world is DCF77, controlled by PTB in Braunschweig, which broadcasts on 77.5 kilohertz. As well as transmitting standard time signals, the broadcast frequency itself is accurate to within five parts in 10^{13} averaged over a ten-day period.

Intervals of time

When it comes to measuring time rather than frequency, we have to distinguish between time intervals and time scales. Time intervals are very important in sports events. In the 100-metre sprint, for example, the best runners can attain speeds of 10 metres per second or faster. Two runners separated by 1 centimetre will cross the finishing line within a mere millisecond of each other, so very accurate timing is needed. But the time of the run is an interval: no one is particularly interested in the UTC at which the runner crossed the line (nor is there any need for an atomic clock!).

There are occasions, though, where time intervals must be measured extremely accurately and only an atomic clock will do. For example, the strength of the Earth's gravitational field can be measured with an instrument called a gravimeter. In principle it is very simple: a small mass is allowed to fall a known distance and the time of the fall is measured with a rubidium clock. In practice the mass falls about 20 centimetres in a fifth of a second and the fall can be timed to a tenth of a nanosecond. That uncertainty corresponds to a distance less than the diameter of a hydrogen atom, and the field strength can be measured to one or two parts in a billion. Again, what matters is the time of the fall and not the UTC.

Sometimes accurate time intervals are needed to measure a distance. Remember the lunar laser reflectors in Chapter 5? The distance to the Moon is measured by timing the journey of a pulse of laser light from the Earth to the Moon and back again. If we know the speed of light we can easily work out the distance travelled by the pulse. At present, distances can be measured to about 3 centimetres, which corresponds to a timing accuracy of 0.2 nanoseconds over a round trip. Using a similar principle the entire gravitational field of the Earth can be mapped by tracking satellites using the laser-ranging methods described in Chapter 4. These depend on accurate measurement of the time of flight of laser pulses, again measured by atomic clocks. This tells geophysicists how mass is distributed within the Earth. Radar works in a similar way, by transmitting pulses of radiowaves which are reflected back from objects, such as aircraft. Timing the round trip of the pulses, which of course fly at the speed of light, tells the operator how far away the aircraft is. A radar required to measure distances to 1.5 metres would need to time pulses to 10 nanoseconds.

A fault on an electricity transmission line may cause protective circuit breakers to trip and cut off the power. It is then necessary to locate and remedy the fault as soon as practicable. Electrical signals travel at very close to the speed of light, so if there is a fault in a power line the control centre finds out about it very quickly by a sudden drop in the voltage on the line. Finding out where the fault is is another matter. Suppose we have a transmission line 30 kilometres long, such that a signal takes 0.1 milliseconds to pass along it. If the break is halfway along the news will arrive simultaneously at each end 50 microseconds later. If the break is closer to one end, then that end will receive the signal an instant before it arrives at the other end. By timing breaks to a microsecond, electricity companies can locate faults on lines to about 300 metres, enough to identify the pylon involved.

The UK Meteorological Office has a system of sensors for determining the position of lightning strikes. A network of radio receivers detects the pulse of radiowaves emitted by a bolt of lightning and by timing the arrival of the pulse at each station the position of the strike can be calculated.

Earthquakes are monitored by seismic stations all over the world. When an earthquake occurs, seismic waves spread through the Earth in all directions. Accurate measurements of the arrival times of the waves at each station allow seismologists to pinpoint the source of the earthquake. A similar method is used to monitor tests of nuclear weapons.

While time intervals are the required measurements in all these cases, it is often more practical to obtain the intervals by first assigning an *epoch* to each event, that is, determining the instant at which it occurred. This is especially true where events are separated by both distance and time. So seismic stations will first record the UTC at which earthquake waves arrive, and the UTCs at many stations are then used to calculate the wave travel times from which the location of the earthquake can be found.

Synchronisation

For many applications we don't just want to know the interval between events but, in effect, the UTC at which they occurred. The absolute time, or epoch, is essential, because different events must be synchronised with each other and with the outside world.

A simple example is the humble railway clock (Figure 7.4). Clocks at thousands of locations on a railway network must read the same time

Figure 7.4. Railway clocks like this one are controlled directly by radio signals from the MSF transmitter.

if trains are to run safely and efficiently. More than that, the railway clocks must also agree with everybody else's clocks if the timetables are to mean anything. How is this to be achieved? The railway company could set up its own master clock, maintained close to UTC, and feed time signals out to the clocks on the stations. But it is far cheaper to invest in station clocks controlled directly by radio signals from a standard UTC time service such as MSF or DCF77. The railway does not need the millisecond accuracy of these services, but having all the clocks independently locked to a common source of guaranteed accuracy saves the railway company the chore of keeping them all synchronised with each other and with the outside world.

Trains do not need to be timed to better than a second, but the same principle is used to meet the much more exacting demands of digital communications networks. Since the 1970s, telephone systems have

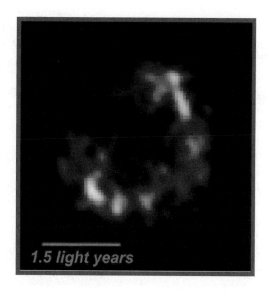

1.5 light years

Figure 7.5. Extremely accurate synchronisation between telescopes is required for the radio astronomy technique of very long baseline interferometry (VLBI). This image of an exploded star in the M82 galaxy, 10 million light years away, was made from data collected by 20 radio telescopes in Europe and North America observing simultaneously in November 1998. The image is 30 times more detailed than can be obtained from the Hubble Space Telescope.

moved over to digital transmission, where signals are sent along cables and microwave links as streams of pulses. This is of course the same method used by computers to talk to each other and the basis of the Internet. For communication to succeed, the streams of pulses must be accurately timed, since the rate at which pulses are launched at one point on the network must match the rate at which they are expected at their destination. And that requires the whole network to be synchronised to a tiny fraction of a second. It is a bit like controlling a railway network: trains are timed so that they do not attempt to run over the same track at the same time. In the same way, trains of pulses are launched onto the network at intervals determined by the clock. The more information is carried on the network, the more accurate the timing needs to be. The International Telecommunication Union recommends that networks should be synchronised to one part in 100 billion, which is equivalent to an accuracy of 1 second in 3000 years. In the UK, BT has recently installed a new timing system for its digital networks. The source of UTC is GPS, backed up by two caesium beam clocks.

Radio astronomers depend on atomic clocks for their most demanding observations. In the technique of very long baseline interferometry (VLBI, introduced in Chapter 5) observations from widely separated telescopes are combined to synthesise a super-large telescope. The key is to record the radio signals on magnetic tape at each observing site and later replay them all together. To do that requires very accurate synchronisation which is provided by hydrogen maser clocks at each site. The timing signal from the clock is recorded at the same time as the radio signal. VLBI can produce detailed maps of astronomical objects as well as monitoring variations in the orientation of the Earth and continental drift (Figure 7.5).

Measuring position

A growing set of applications of accurate time are actually concerned with measuring positions on the Earth using the navigational satellite systems GPS and GLONASS. These require a specialised receiver which detects the signals from the satellites in view (usually five or more) and automatically computes the latitude, longitude and height of the receiver. Even the simplest GPS receivers can pinpoint horizontal positions on or near the Earth to better than a hundred metres, and this precision is utterly dependent on the stability and accuracy of the GPS time scale.

Better performance is possible using "differential" techniques. Here a number of fixed reference stations of known position monitor the GPS signals and issue corrections that can be used by GPS receivers in the same area. Differential methods can circumvent the degradation imposed on the GPS signals intended for civilian use (now removed), and accuracies of a few metres are possible. With even more sophisticated methods of processing signals, positions can be pinpointed to a few centimetres.

Although GPS was developed for military purposes, its civil applications now far exceed anything that was dreamed of when the service began. Aside from the obvious uses—location of ships, aircraft and spacecraft—GPS location is now being used for many purposes where it has replaced less convenient methods of position finding.

For example, if you are not much good at reading road atlases on the move, you can now buy a unit that displays the position of your car on an electronic map. If your car (or perhaps the electronic map inside) is valuable enough you can equip it with a discreet black box that will compute and radio its position should the car be stolen. Similar devices are being marketed to replace lorry tachographs and even taxi meters. Operators of emergency services can use GPS to keep track of their vehicles and despatch the nearest to the scene of an incident. Displays at bus stops can tell passengers the location of the next bus and how long they will have to wait. Trains can be tracked through sparsely populated areas. Hikers can use hand-held receivers to guard against getting lost in the mountains. There is even a proposal for an "electronic guide dog", a small computer that would give continuous voice directions to blind people, based on its location determined by GPS: "Stop here, the post office is on the right."

All of these are inventive extensions of the primary use of GPS, but there are others which are not so obvious. We saw in Chapter 5 how GPS can be used to monitor the orientation of the Earth itself, and so track small changes in the length of the day and polar wobble. It can be used to measure the height of the oceans or movements of the Earth's crust in earthquake zones. Many uses of GPS can be classed as surveying. It can be used to determine the location of boundaries in featureless landscapes, assist in hydrographic surveys and mineral prospecting, and even set out lines on sports fields. Farmers can use GPS to map out large fields and control the application of fertilisers on the scale of a few metres rather than apply it indiscriminately to the whole area. In the corn belt of North America, GPS-controlled robotic harvesters tirelessly roam the prairies gathering crops while the farmer stays at home.

There appears to be no limit in sight to this particular use of accurate time.

New units for old

This book has been about the measurement of time and its unit, the second. But the second is only one of seven "base units" that make up the Système International (SI). The others are the metre, the unit of length; the kilogram, the unit of mass; the ampère, the unit of electric current; the kelvin, the unit of temperature; the candela, the unit of luminous intensity; and the mole, the unit of "amount of substance". Suitable combinations of these base units allow us to construct other units to measure any physical quantity. Figure 7.6 shows how the base units are related to each other and how accurately they can be realised in practice. Of the seven units, the second stands out as the most accurately known, which is why time can be measured with such precision. In contrast to time, temperature can only be measured to three parts in ten million and luminous intensity to only one part in 10 000. Over the years there have been attempts to relate other base units to the second and so further exploit the potential of this great accuracy.

Each of the base units has its own precise definition, which we won't go into here, but three of them already depend on the definition of the second, namely the metre, the ampère and the candela, though the latter two also depend on the definition of the kilogram. The metre, however, is uniquely dependent on the second.

In Chapter 1 we saw that the metre used to be defined as the distance between two scratches on a metal bar kept at BIPM in Paris. Various measurements over the years showed that the metre bar was stable to three parts in ten million, but by the 1950s measurement techniques had already surpassed that accuracy. In much the same spirit that eventually saw the second defined in terms of caesium, the CGPM redefined the metre in 1960 in terms of the wavelength of a particular emission from a particular atom, namely 1650 763.73 wavelengths in vacuum of the orange–red line in the spectrum of the krypton-86 atom.

In 1983 it was redefined again in a most interesting way. Measurements of the speed of light had become so good that they were potentially more accurate than the metre itself. Indeed a measurement from 1972, of 299 792 458 metres per second, was accurate to four parts in a billion, and this accuracy was limited chiefly by the difficulty of using the krypton lamp as a length standard. The decision was then

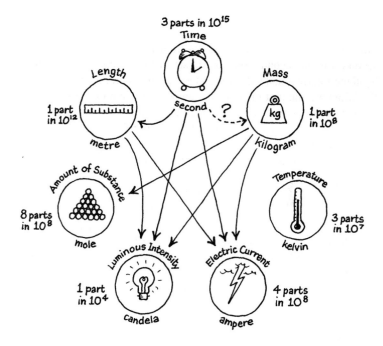

Figure 7.6. The seven base units of the SI: how accurately they can be realised and how they are related to each other. The question mark shows a possible redefinition of the kilogram in terms of the second. With that link in place, six of the seven units would ultimately be defined in terms of the vibration frequency of the caesium atom.

taken to define the metre in terms of the speed of light. So since 1983 the metre is "the length of the path travelled by light in vacuum during a time interval of 1/299 792 458 of a second". This has the curious consequence that it is no longer possible to measure the speed of light: it is fixed forever at exactly 299 792 458 metres per second.

More importantly, the 1983 definition ensures that the metre—in principle at any rate—can be realised with the same accuracy as the second. As the uncertainty in the duration of the second edges down towards one part in 10^{15}, only technological limitations prevent the metre being realised to the same accuracy. In fact, as we shall see in the next chapter, distances can now be measured in practice to a few parts in 10^{11}, about 100 times better than two decades ago.

The success of this elegant new definition has led metrologists to ask whether the third of the traditional base units—the kilogram—could also be defined in terms of fundamental constants. At the moment the kilogram is defined as the mass of the international prototype kilogram, a cylinder of platinum–iridium kept at BIPM. Replica kilograms, the national prototypes, are kept at national standards laboratories (Figure 1.8). It is not a very satisfactory definition, since it depends not on the constants of nature but on the mass of an artefact, a lump of metal, which may change over the years. Indeed, measurements of the national prototypes suggest that they are gaining mass at the rate of about 1 microgram each year, presumably from atmospheric contamination. The last time the replicas were compared with the BIPM standard, in 1993, their masses covered a range of about 100 micrograms. For these reasons, the present definition of the kilogram is regarded as satisfactory only down to accuracies of about one part in 100 million.

Two alternative approaches to a new definition are being investigated. Each of them requires one of the fundamental physical constants to be fixed, in the same way as the speed of light was fixed to define the metre. One idea is to define the kilogram as a certain number of specified atoms, usually atoms of carbon. One suggestion is: "The kilogram is the mass of 5.018×10^{25} free carbon-12 atoms at rest and in their ground state." In effect, this fixes the Avogadro constant (sometimes called Avogadro's number) which relates the number of atoms in an object to its mass. Its accepted value is known to about six parts in 10 million, but this figure would need improving if the kilogram were to be redefined in this way. At the moment several collaborating laboratories are using state-of-the-art techniques to count the number of atoms in a specially made sphere of silicon, constructed to have almost the same mass as the international prototype kilogram. It is a bit like the competition at school fetes to count the number of sweets in a jar, except that no one knows how many are really in there.

The second approach, which may prove more practical, is to define the kilogram in terms of a frequency. Yet another consequence of

Figure 7.7. This apparatus, known as a Watt balance, is being used at NPL in a series of experiments that may ultimately lead to the kilogram being redefined in terms of the caesium frequency.

Einstein's relativity is that energy possesses mass. This is the famous $E = mc^2$, which tells us the energy E associated with a mass m, where c is the speed of light. Back in Chapter 2 we saw that a photon of light possesses an energy directly proportional to its frequency. So x-ray photons, for example, are much more energetic than photons of visible light. It follows that a photon of a specified frequency will have a particular energy which will in turn possess a particular mass. We could then define the kilogram as the mass of a certain number of photons of a certain frequency. In fact, one need not even do that, since all that matters is the total frequency of the photons. Since frequency is the number of cycles in 1 second, we would be defining the kilogram in terms of the second. One definition that has been proposed is: "The kilogram is the mass of a body at rest whose equivalent energy equals the energy of a collection of photons whose frequencies sum to $135\ 639\ 274 \times 10^{42}$ hertz."

In this case the constant that is being fixed is Planck's constant, which relates the frequency of a photon to its energy. Planck's constant is the fundamental quantity of quantum physics, and its accepted value is known to about one part in 10 million. Experiments underway at NIST and NPL are expected to improve this uncertainty to around one part in a billion (Figure 7.7), and in that case the new definition might become very attractive.

If metrologists go down the frequency route to redefining the kilogram, then all three of the traditional base units will ultimately be defined in terms of the caesium transition. As can be seen in Figure 7.6, these three uniquely define three of the other units, the ampere, the mole and the candela, with the potential for greatly improving the accuracy of each. In that case, only the kelvin, the unit of temperature, will remain aloof from the pervasive influence of time.

8

THE FUTURE OF TIME

The Long Now

Where does time go from here? It depends how long a view you want to take. The longest view of all is being taken by the Long Now Foundation, a group of enthusiasts who believe that our preoccupation with the here and now is causing us to lose sight of the scale of time. They want to widen our perspectives by constructing a monumental clock that will last for 10 000 years. Danny Hillis, the US computer scientist who is the inspiration behind the project, imagines a clock that "ticks once a year, bongs once a century, and the cuckoo comes out once every millennium".

Since no one can forecast how human civilisation will fare over 10 millennia, the clock's designers are making few assumptions about the technological abilities of our descendants. One of the ground rules is that the clock must be repairable with bronze-age technology, which rather rules out caesium beams. The clock's workings will be entirely mechanical, using a mechanical oscillator as a frequency standard and a mechanical computer to count the cycles (Figure 8.1). It will automatically keep itself in check by periodic calibrations from the noon Sun.

With an accuracy of one day in 2000 years the clock will hardly compete with atomic clocks, but that is not the point. The Clock of the Long Now will still be ticking long after the atomic clocks at NPL are being transformed into interesting fossils in the sediments of the Thames basin.

In this chapter we shall take a distressingly shorter view of time. If we confine ourselves to the next couple of decades, the successors to

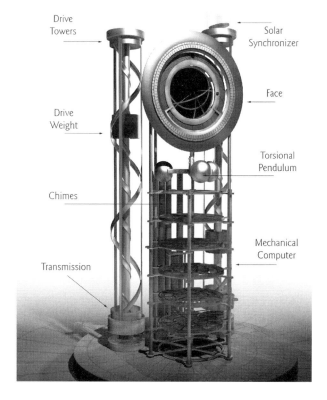

Figure 8.1. One of the proposed designs for the Clock of the Long Now, intended to run for 10 000 years. A small-scale prototype was finished in time to strike the third millennium.

the caesium beam standards, the workhorses of atomic time for more than 40 years, are already waiting in the wings. They in turn will be succeeded by even more radical technologies that are now being nurtured in laboratories around the world. And there may even be changes in store for UTC itself. But first we will look at the future of the primary standards.

Caesium fountains

We have already seen how the world's first caesium fountain, now work-ing in Paris, is the first of a new generation of machines that will of-fer perhaps ten to a hundred times improvement in performance over existing caesium beam standards. Almost every laboratory operating a primary caesium standard is developing a caesium fountain and several others have joined in as well. A second fountain is now working at NIST and in the next few years caesium fountains may be operating in eight to ten centres around the world, providing a badly needed increase in the number of primary standards.

One of the laboratories working on caesium fountains is NPL, where Louis Essen and Jack Parry constructed the first working caesium standard in 1955. An experimental caesium fountain is under test in a former beer cellar at Bushy House, the seventeenth century mansion that was the original NPL building (Figure 8.2). In the early years of the twentieth century the room had been used for testing watches and chronometers, so perhaps it is appropriate that it should now house the timepieces of the twenty-first century.

You may remember from Chapter 4 that atoms in a caesium foun-tain are gently lobbed vertically to a height of a metre or so before falling back down. Once on the way up and again on the way down they pass through a cavity where they are exposed to microwaves. If the microwaves are tuned to the transition frequency the atoms are flipped from one hyperfine state to the other. The key to the technique is to obtain a supply of sufficiently slow atoms. If you want the atoms to rise to a height of about a metre you need to toss them up with a speed of 4.5 metres per second but, as we shall see, they need to be slowed to a few centimetres per second before this can be done. Caesium atoms in a con-ventional beam standard emerge from the oven at around 200 metres per second, so how do you get slow atoms? How that happens is the key not only to the caesium fountain but also to other new types of atomic clock.

In Chapter 2 we saw that an atom will absorb a photon of light, provided that the photon has precisely the right energy to lift an electron inside the atom from one level to a higher level. Caesium will absorb photons of a wavelength of 852 nanometres, which is near-infrared radi-ation, invisible to the eye. One of the technological innovations which

Figure 8.2. Peter Whibberley (left) and Dale Henderson at work on the experimental caesium fountain at NPL. The fountain itself is contained within the rectangular framework above centre, while the optics to the bottom left are used to prepare the laser beams. The laboratory was once used for calibrating marine chronometers.

made this work feasible was the invention of reliable lasers operating at this wavelength—the same kind of lasers that are used in CD players.

If a caesium atom is exposed to light of wavelength 852 nanometres, it will absorb a photon and almost immediately re-emit it again, as if the photon has bounced off the atom. Indeed, this process is known as

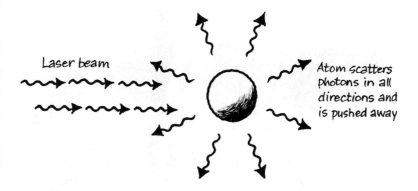

Figure 8.3. A caesium atom will scatter photons from a laser tuned to the correct wavelength and be pushed along by the recoil.

0 "scattering" of light. When the atom absorbs the photon it receives a little kick of momentum in the direction the photon was travelling. When it re-emits the photon the atom recoils with another little kick in the direction opposite to the photon's travel. At first sight one might think that these kicks would cancel out: for every incoming photon there is an outgoing photon too. But if the atom is in a laser beam, the absorbed photons all come from the same direction while the scattered photons are sprayed out at random. The kicks do not balance out and the atom in the beam gets pushed along by the light, scattering photons as it goes (Figure 8.3).

Now suppose the atom is moving towards the source of the laser light. Does it absorb photons still? No, because although the photons are the right wavelength for a stationary atom, the Doppler effect (Chapter 2) ensures that the moving atom sees them blue-shifted to a slightly shorter wavelength. The atom sees photons streaming past that are too short to be absorbed and nothing happens. But if we now adjust the wavelength from the laser to make it slightly *longer* than 852 nanometres, then provided we get it just right the moving atom will see these longer photons blue-shifted to 852 nanometres and begin to absorb them! This time, because the atom is moving into the beam, the little kicks act against the motion and the atom slows down. A caesium atom in a vapour at room temperature can be slowed almost to a standstill by absorbing and re-emitting about 70 000 photons. Because the temperature of a gas is just a measure of how fast the atoms are moving, an atom which has been slowed in this way can be said to be "cooled". This technique is known as "Doppler cooling".

So important has this principle become in many areas of physics, that the scientists who first developed this and other techniques of laser cooling—Steven Chu, Claude Cohen-Tannoudji and William Phillips—shared the Nobel Prize for Physics in 1997.

In the NPL experimental fountain, solid caesium sits in a reservoir attached to the bottom of a vacuum chamber (Figure 8.4). It simply evaporates from there, filling the chamber with a thin atmosphere of caesium atoms. The vacuum—about a thousand billion times less than atmospheric pressure—ensures that stray molecules of air will not impede the movement of the atoms.

In their random walk around the chamber the atoms will sooner or later be caught in the intersection of six laser beams, all of wavelength slightly greater than 852 nanometres. If the atom remains stationary, nothing happens. But if the atom moves in any direction, it will be moving towards one or more of the beams and see blue-shifted photons. It will then start to scatter them and be slowed down.

Many millions of atoms can be collected and cooled at the inter-section of the beams. Because they meet resistance whichever way they move, the region where the atoms are held has been called "optical molasses". Atoms constantly move in and out of the molasses, becoming trapped for a short time before escaping.

The amount of cooling is extraordinary. Normal everyday temper-atures are about 300 degrees above absolute zero, or 300 kelvin (written 300 K). Water freezes at 273 K and the coldest temperature recorded anywhere on Earth is about 215 K. At 77 K the molecules of atmospheric nitrogen—the main component of the air—move so slowly they turn into a liquid and at 63 K they freeze. Doppler cooling can cool caesium atoms to much less than 1 K.

Theoretical calculations showed that the minimum temperature for Doppler cooling of caesium is an extraordinary 0.000 125 K, or 125

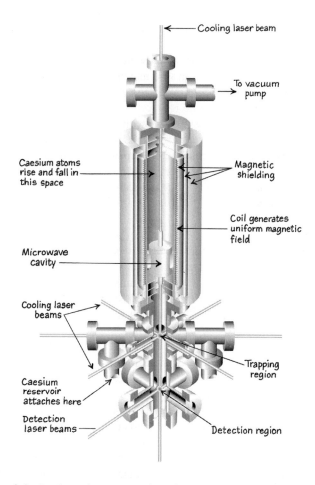

Figure 8.4. A schematic representation of the experimental caesium fountain developed at NPL.

microkelvin. At these extremely low temperatures atoms are moving at speeds of only a few centimetres per second. But physicists have now discovered and refined other cooling mechanisms which can achieve even lower temperatures. In a caesium fountain the atoms are routinely cooled to a few microkelvin, and more elaborate techniques have cooled atoms to a few nanokelvin—billionths of a degree.

In the NPL caesium fountain the atoms collect in an ultra-cold cloud in the optical molasses at the intersection of the laser beams. To hold them there a shaped magnetic field is applied, forming a "magneto-optical trap". There is no need to inject them into the molasses or anything so sophisticated: they stumble into the trap and are caught.

When the cloud is ready for its flight through the chamber the trapping magnetic field is turned off, then slight changes in the wavelengths of the vertical laser beams propel them upwards. As the atoms rise towards the cavity, a short burst of microwaves flips the atoms into the lower hyperfine state, and a final flash from a laser blows away any that remain behind.

Now with all the lasers turned off the caesium atoms rise in darkness through the radiation in the microwave cavity, emerge from the top of the cavity, reach a standstill then fall back through the cavity. Provided the frequency is correct, the two exposures to the microwaves are enough to flip most of the atoms into the upper hyperfine state. The atoms fall back through the trapping region—now inactive—towards the bottom of the chamber. There, more laser beams excite the atoms and the number in each state is counted by the photons they re-emit. As with other caesium standards, the frequency of the microwaves is tuned until the proportion of atoms making the transition is at a maximum. Then the microwave frequency matches the magic number of 9192 631 770 hertz.

One would like the atoms to spend as long as possible in flight since the longer the time between the microwave exposures the sharper the clock frequency will be. But as soon as the cloud of atoms is lofted from the trap it starts to expand, and many of the atoms end up on the walls of the vacuum chamber rather than falling back through the microwave cavity. This is the chief reason for cooling the atoms as much as possible before launching them. In practice, clouds are launched at intervals of about 1.2 seconds.

The biggest problem remaining to be solved is a rather weird effect due to quantum mechanics: atoms moving very slowly get bigger. A caesium atom moving at 1 centimetre a second appears about 500 times its normal diameter. Not only do these bloated atoms collide so often that the precision of the device becomes blunted, but the energy levels become distorted as well.

Workers at NPL are constructing a second caesium fountain which will be operated as a primary standard. Several others should come on-line in the next few years, joining the pioneering machine at LPTF in Paris. These machines will greatly improve the accuracy of TAI and UTC, but the conventional caesium beams will still have a part to play for many years to come.

Clocks in space

Caesium fountains are attractive because the interaction time of the atoms can be as long as a second, as they rise and fall through the microwave cavity. But suppose it were possible to operate a fountain in an orbiting spacecraft, where there was no gravity to pull the atoms down again. Once a group of atoms has been cooled by laser—in the same way as in an earthbound caesium fountain—the atoms will remain at their slow speeds even when the confining lasers are turned off. On Earth, they would immediately drop under gravity, but in space we could, in principle, produce a bunch of supercold atoms and then let them float as long as we wish while exposing them to microwaves.

The idea is being taken very seriously and at present there are two proposals to put experimental atomic clocks on the International Space Station now being constructed 400 kilometres above our heads. One of these is from the US and the other is from Europe.

The European project is called ACES (Atomic Clock Ensemble in Space) and is scheduled to fly on the International Space Station from 2002. The principal instrument will be PHARAO (Projet d'Horloge Atomique par Refroidissement d'Atomes en Orbite), a caesium atomic clock based on the design of the LPTF caesium fountain. In conditions of microgravity the clock is expected to attain a stability of one part in 10^{16} to 10^{17} in one day, with an accuracy of one part in 10^{16}. The second clock will be the Space Hydrogen Maser (SHM) contributed by the Observatory of Neuchâtel. Its prime purpose will be to act as a local reference standard for evaluating the performance of PHARAO.

In Chapter 5 we saw how a network of tracking stations uses laser beams to measure the distances to orbiting satellites such as Lageos.

Some of those stations will be used in a novel technique for transferring time from ACES. Mounted on the outside of the space station, ACES will carry an array of laser reflectors similar to those placed on the Moon and on the Lageos satellites. ACES will time the arrival of laser pulses by its on-board atomic time scale, while the ground station computes the time delay between transmitting a pulse and receiving the returned flash. This variation on the two-way transfer method will allow ACES time to be transferred to BIPM for inclusion in TAI. Because laser transfer cannot work in cloudy weather, a microwave time transfer link will also be available.

Both ACES and the US project are experimental, and it is by no means clear that the future lies in space. There are problems with operating an atomic clock on a busy space station, such as having to contend with noise and vibration from the other on-board activities, as well as local sources of magnetic fields.

While ACES is a test project for greater things, it should allow TAI and UTC to be transferred with an accuracy 100 times better than is currently achieved and will be much in demand for new tests of general relativity.

After caesium

The 1967 definition of the second has proved its worth and the caesium atom is still the element of choice for constructing atomic clocks. But it may not always remain so.

Several groups around the world are investigating whether frequency standards can be constructed that operate in the optical part of the spectrum; that is, with visible or near-infrared light. The advantage of going to optical wavelengths is that the frequencies are so much higher. Where a caesium clock using microwaves "ticks" at around 9 billion times a second, a clock using visible light would tick about 70 000 times faster, allowing time intervals to be measured with a similar increase in precision.

Timekeepers are not the only scientists interested in optical frequency standards. Because the metre is now defined in terms of the second, as we saw in the last chapter, a frequency standard is equivalent to a length standard. In practice it is not as simple as that, since precise length measurements are made with lasers operating with visible light, while caesium standards use microwaves. The best length standards at present are iodine-stabilised helium–neon lasers emitting red light of wavelength 633 nanometres. Although these standards are accurate to 2.5 parts in 10^{11}, length metrologists would welcome an even better primary standard in the optical spectrum against which all such lasers could be compared.

Catching an atom

It has been called the spectroscopist's dream. The ideal frequency standard would be a single atom at rest in free space, undisturbed by external influences. In Chapter 2 we saw how Lord Kelvin, more than a century ago, had speculated that atoms could become "natural standards" of length and time by virtue of the precise wavelengths of light they emit and absorb. Several research groups around the world are now attempting to devise frequency standards based on the light emitted and absorbed, not just by atoms but by *single* atoms.

But which atom? It turns out it is much easier to trap an *ion* rather than an atom. An ion is an atom that has lost or gained one or more electrons, so that the positive charge on the nucleus is no longer balanced by an equal number of negative charges. An ion therefore carries a net electric charge, which means it can be caught and held in an electric field, something that is not possible with uncharged atoms. So the idea is to find an atom that will have a convenient arrangement of electrons after one has been removed. Many atoms have been looked at, especially the alkaline earth metals, which all have two outer electrons. Removing one of these leaves a single outer electron, which gives the ion a particularly simple structure—much like caesium, in fact. These include beryllium, magnesium, calcium, strontium and barium. Other promising elements include mercury, indium and ytterbium.

Several groups have been operating experimental frequency standards based on trapped ions for years. But these work at microwave frequencies, using similar kinds of hyperfine transitions to those found in caesium. A group at NIST in the US has made a frequency standard by trapping a handful of mercury ions (Figure 8.5). As in the caesium

Figure 8.5. A string of mercury ions is held in a trap at the National Institute for Standards and Technology. Such ions can be used as frequency standards by exposing them to microwave radiation at the hyperfine transition frequency.

fountain, the ions are cooled and pumped by lasers, but instead of being launched through a microwave cavity the ions remain in the trap where they are exposed to two bursts of microwaves as long as 100 seconds apart.

While this has the potential for great accuracy, the real prize is to learn how to make standards that work at optical frequencies. Research groups in the US, Canada, Germany, France, UK, Japan and China are all working towards this goal. A number of other countries are also using related ideas to make ion-based clocks.

One group at NPL, for example, has built a trap designed to isolate a single ion of strontium, a soft metal which is used to give the red colour to fireworks and distress flares. Similar work is being pursued at the National Research Council of Canada. Atoms are evaporated from a small oven, ionised by a hot filament, and then captured by an electric field. So delicate is the technique that the NPL physicists can usually trap as few as one or two ions in a 1-minute loading sequence. If more than one is trapped they start again. A laser cools the ion to less than a millikelvin by a similar Doppler cooling technique to that used in the caesium fountain. Once cooled, an ion remains imprisoned indefinitely, or more likely until a stray molecule bumps into it. The record at NPL for keeping a solitary ion is more than one month (Figure 8.6).

Once a single ion is trapped and cooled, the next step is to locate a suitable "clock" transition which could provide a standard frequency analogous to the caesium spin–flip transition. The best defined transitions, those with the sharpest frequencies, are also those with the longest lifetimes. When an electron is excited up to a higher

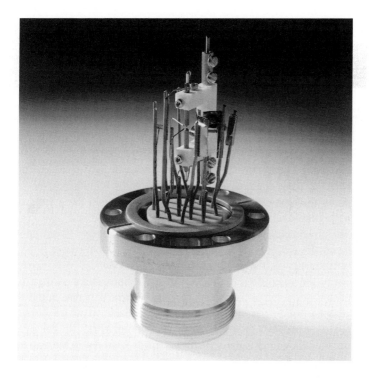

Figure 8.6. A trap constructed at NPL to confine ions of strontium or ytterbium. The ion is trapped between the two pointed electrodes above the centre of the picture.

energy level it waits a short time before dropping down again. Mostly this time is very short, only a few nanoseconds, but the longer an electron waits before falling back to the ground state, the sharper will be the photon frequency. And that means that the longer-lived states are ideal for future frequency standards. The clock transition in strontium has a wavelength of 674 nanometres and a lifetime of about one-third of a second. That means that the strontium ion could emit an average of three photons a second at that wavelength.

But if the transition is that weak, how can it be detected? The answer is a technique proposed by the German physicist Hans Dehmelt, for which he was awarded a share of the Nobel Prize for Physics in 1989, known as "quantum jumps" or "electron shelving". The strontium ion, held in a trap, is illuminated with the cooling laser operating at the blue wavelength of 422 nanometres (Figure 8.7). As with the caesium atoms described earlier, the strontium ion continually scatters the laser photons, spraying them off in all directions at a rate of a hundred million a second. The single ion can be seen as a tiny point of blue light in the middle of the trap.

A second laser now illuminates the ion with light near 674 nanometres. The wavelength of the laser is adjusted by minute amounts until it hits on the correct value to excite the electron from the ground state to the higher level. At that point the electron is "shelved" and the ion can no longer scatter the photons from the cooling laser—it disappears! So when the ion goes dark, you know that the probing laser is on the same frequency as the clock transition. It is identical in principle to adjusting the microwave frequency of a caesium clock to match the frequency of the hyperfine transition. The electron remains on the shelf for about a third of a second before falling back, at which point the ion starts scattering again and the blue dot reappears. So the sign of hitting the clock frequency is when the steady blue dot in the trap turns into a flickering point of light. This flickering, incidentally, is clinching evidence that the blue dot is a single ion.

Much the same techniques are being used around the world to investigate other ions. While it is not yet clear which ion will prove to be the best for a clock, there is a lot of interest in ions which have unusually sharp transitions. In 1997 physicists at NPL were the first to excite a very weak transition in ytterbium, a rare metal that occurs in minute quantities in the Earth's crust. (The same ion is also being investigated by a group at PTB in Braunschweig, though at a different wavelength.) They estimated the transition to have a lifetime of no less than ten years, the longest ever discovered. Once the electron is placed in that state it would be stuck there for a decade before it spontaneously fell down, emitting a photon of wavelength 467 nanometres. Since no one is going to wait ten years for the electron to come down off the shelf, a third laser can be used to lift it to an even higher level from where it drops quickly

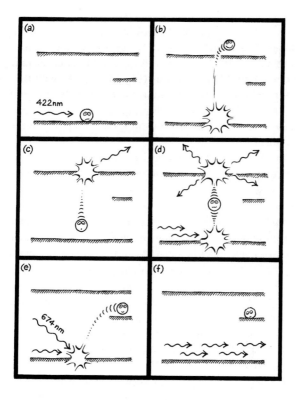

Figure 8.7. Finding the clock transition in a strontium atom. (a) The electron remains in the ground state until illuminated with blue light of wavelength 422 nanometres. (b) It absorbs a photon and is excited to a higher energy level in the atom. (c) Within a few nanoseconds the electron falls back, emitting an identical photon and the cycle can repeat. (d) With constant illumination the atom is seen to scatter blue light. (e) Now the atom is probed with a second laser of wavelength near 674 nanometres. If the wavelength is correct, the electron absorbs the photon and is excited to a third state where it remains for about a third of a second. (f) While on the "shelf" the electron cannot scatter the blue photons and the atom goes dark, indicating that the probing laser has found the clock transition.

to the ground state, and the process can start all over again. Because of its long lifetime, ytterbium has the potential for providing an extremely stable frequency standard.

Another possibility is indium, a silvery white metal much used in the electronics industry. Researchers at the Max Planck Institute for Quantum Optics near Munich and also at the University of Washington in Seattle are investigating a transition at 237 nanometres which, though not as narrow as ytterbium, is easier to control and may turn out to be more practicable.

Alongside the search for promising transitions, much work is going on to develop the supporting technology. If very narrow transitions are to be probed with sufficient finesse to be useful, then the illuminating laser must also have a very sharply defined wavelength. The pioneering group at NIST, who already have a working microwave standard based on mercury ions, have built a laser several hundred times more stable than those available elsewhere. They plan to use it to investigate the narrow optical transition in mercury at the ultraviolet wavelength of 282 nanometres.

The chief obstacle to turning these advances into a working clock is that the frequencies are very high—this is, of course, the same reason that an optical clock would be so desirable. The ytterbium wavelength of 467 nanometres, for example, corresponds to a frequency of 6.4×10^{14} hertz, and it is not yet possible to count directly the cycles of something that vibrates 640 trillion times a second. The best that could be done in the near future is to relate the frequency to a caesium standard by a number of intermediate steps, but the accuracy would be set by the caesium clock, not the optical transition.

If atomic clocks based on optical transitions do come along, and it would be many years in the future, researchers expect to attain stabilities of around one part in 10^{18}. That means that such a clock, set running at the instant of the Big Bang, would today still be within 1 second of the right time.

Better UTC

Better primary standards and secondary clocks will mean that the accuracy and stability of TAI and UTC continue to improve, as they have

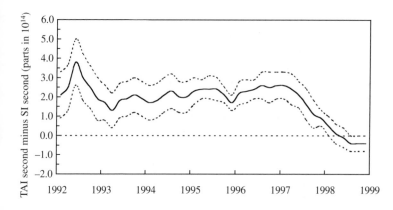

Figure 8.8. The second of TAI—and of UTC—has been kept to within a few parts in 10^{14} of the SI second in the 1990s. The dashed lines show the limits of uncertainty in the estimated accuracy.

steadily done over the past few years. In 1998 the accuracy of the TAI second was estimated to be within five parts in 10^{15} of the SI second (see Figure 8.8), while the stability of TAI over a standard period of 40 days attained one part in 10^{15} for the first time.

There are stirrings, though, about the famous leap second. In the spring of 1999 timekeeping professionals from around the world met in Paris for one of their regular conferences. Dennis McCarthy, Director of Time at the US Naval Observatory, raised a series of questions about the future of the leap second. Since the present version of UTC was launched in 1972, leap seconds have been introduced to ensure that UTC is maintained close to UT1, the time kept by the rotating Earth. This provision was made to assist navigators, who needed a time that bore a close relation to the orientation of the Earth in space. With a maximum error of 0.9 seconds, anyone using UTC time signals in conjunction with star sightings would find their position in error by no more than 420 metres.

But, as McCarthy pointed out, no one navigates by the stars any more. Virtually all commercial navigation by air and sea is carried out using the GPS and GLONASS systems and other electronic aids. In fact, leap seconds are increasingly more of a hindrance than a help. Some computer systems cannot cope with leap seconds at all, and have to be reset when they occur. If the present system were to continue unchanged, McCarthy said, we should expect about three leap seconds every two years by the year 2050. So do we really need leap seconds at all?

One possibility is to reduce the number of leaps by increasing the permitted deviation from UT1. If the difference were allowed to grow to, say, 4.9 seconds, then a leap of 5 seconds could be introduced at less frequent intervals. There would be fewer leaps but the leaps would be bigger. Another idea is to save up the leap seconds and introduce several of them at once on designated dates. The size of the leaps would not be predictable but at least the dates would be known in advance.

Back in Chapter 5 we saw how the number of leap seconds depends more on the definition of the second than on any changes in the rotation of the Earth. So the most radical solution would be to redefine the second to match more closely the present length of the day, so reducing the number of leap seconds needed. But as the second underpins the entire international system of units, this would have knock-on effects far beyond timekeeping. And tidal drag on the rotating Earth would ensure that the ideal second today is no longer ideal a few decades hence.

Or perhaps we should simply abandon leap seconds altogether and allow UTC and UT1 to drift apart. If we did this today, the difference would amount to about a minute by 2050. Those who needed to use UT1 would have access to the necessary corrections and the public would not start to notice any effect on the timing of day and night for the foreseeable future.

There was no enthusiasm at the meeting for reopening the question of leap seconds and UTC, but clearly there is some underlying discontent which is bound to surface again in the new century.

There are likely to be changes, though, in the way that UTC is computed and disseminated. One of the peculiarities of UTC is that it is only available a month or so in arrears. All the time signals disseminating "UTC" are actually broadcasting approximations to UTC made by one or other of the national timing centres. Only after several weeks do the centres find out from BIPM how good their approximations were.

As demand for accurate time grows, the challenge of computing and disseminating UTC more quickly is likely to grow along with it.

As far as the dissemination of time is concerned, navigational satellites will have it largely to themselves. With GPS time readily available anywhere in the world to within a few tens of nanoseconds of UTC, there seems little point in deriving accurate time from any other source. This poses a new challenge to the national timing centres, who find their quasi-legal roles as the sources of "national" time being bypassed by GPS. But where does GPS time come from? It comes in the first instance from the version of UTC maintained by the US Naval Observatory, but that in turn is validated through BIPM by the collective efforts of dozens of national centres.

While there may be fewer customers for locally disseminated UTC, the national laboratories will continue to underwrite the stability of the world's timekeeping system through their ensemble of numerous secondary clocks, and a small but increasing number will invest in the primary standards that guarantee the accuracy of UTC no matter how it is disseminated. Another, and perhaps growing, role for national laboratories is as independent third parties who can validate time and frequency measurements and advise their national industries and public services. And let us not forget the network of observatories and tracking stations who, through IERS, monitor the rotation of the Earth and ensure that UTC remains close to the everyday time kept by the rising and setting Sun. The beauty of the world timekeeping system is that it is decentralised and dispersed and therefore robust; it is not open to abuse by any one nation or organisation.

That said, there is some concern that so much of the civilian economy is growing to depend on time disseminated through GPS, a navigation system operated by the US military with no international accountability. Is it healthy for so much of the world to rely on GPS? The emergence of GLONASS as an alternative offers some reassurance, but that too is a military project and the continued political and economic uncertainty in Russia, not to speak of the poor state of the satellites themselves, will deter people from relying too heavily on GLONASS for critical applications.

For that reason there are moves in Europe to create an entirely civilian satellite navigation system. The first stage—known as EGNOS—is a European geostationary satellite that will work alongside GPS by relaying information about the status of GPS satellites gleaned from a network of monitoring stations. The idea is that users will learn about faulty satellites much faster than they do now, an advance that will be particularly welcome to the aviation industry. Further ahead, the European Union is studying plans to create its own navigation system, known as Galileo, which will be similar in principle and scale to GPS and GLONASS but including a small number of satellites in geostationary orbit in permanent view from Europe.

Could it have been different?

In October 1998 the Royal Greenwich Observatory closed its doors for the last time. It had been founded by King Charles II in 1675 to provide services to navigators, notably by supplying astronomical information to enable them to find longitude at sea. This essentially meant finding *time* at sea, and the provision of accurate time remained a core activity of the observatory until the 1980s.

By the 1990s RGO no longer had any responsibilities for time, and had long diversified into the provision of observing facilities for university research groups as well as conducting its own astronomical investigations. The UK government took the view that the observatory was doing nothing that could not be done by other institutions and 323 years of history came to an abrupt end.

One theme of this book has been the radical changes in timekeeping through the twentieth century, in which responsibilities for time have passed decisively from astronomers to physicists. It is notable that the surviving national observatories, such as the Paris Observatory and the US Naval Observatory, are deeply committed to atomic timekeeping. The Paris Observatory hosts the IERS as well as the world's most accurate atomic clock, and the formidable ensemble of clocks at USNO contributes about 40 percent of the weighting of TAI.

But the story could have had a different ending. In the last chapter we met the pulsars, rapidly spinning magnetic stars that pour out streams of highly regular pulses. Today, more than 1000 pulsars have been discovered and their pulse periods are known with great precision. For example, the rotation period of the fastest pulsar, PSR 1937+21, is catalogued as 0.001 557 806 492 4327 seconds, with an uncertainty of plus or minus four in the last decimal place. That's equivalent to an accuracy of 2.6 parts in 10^{13}, comparable with the primary frequency standards listed in Chapter 4.

In the 1980s astronomers were speculating that pulsars could provide a more stable time scale than atomic clocks. Although pulsars are gradually slowing down, the rate of slowing is also well known and an ensemble of the most regular pulsars would provide an accessible time scale of great stability. But science moves on, and at the turn of the century it is hard to see how a collection of spinning stars could compete with the elegance and precision of a time scale based on the fundamental constants of nature as realised by caesium fountains and ultimately by the radiation of a single atom floating free in space.

Pulsars were discovered in 1967 but radio astronomers had the means to detect them a decade earlier. One wonders what now would be the fate of astronomical time if pulsars had been discovered in the 1950s. If the IAU had chosen to replace Universal Time with Pulsar Time instead of the esoteric and inaccessible Ephemeris Time our story might have been different.

Appendix

TIMEKEEPING ORGANISATIONS

International organisations

Several organisations share the responsibility for the world timekeeping system. They all have websites, some of which contain extensive resources on timekeeping.

Bureau International des Poids et Mesures (BIPM)

BIPM is custodian of all the world's units of measurement, and since 1988 has been responsible for the formation of the TAI and UTC time scales. The Bureau's monthly Circular T, which publishes corrections to UTC disseminated by the national timing centres, can be downloaded from the website.

Address: Time Section, BIPM, Pavillon de Breteuil, F-92312 Sèvres, France.
Website: http://www.bipm.fr

International Earth Rotation Service (IERS)

Created in 1988 by the International Astronomical Union (IAU) and International Union of Geodesy and Geophysics (IUGG), the IERS maintains terrestrial and celestial reference frames and monitors the Earth's rotation. It is responsible for deciding when a leap second is required to maintain UTC close to UT1. The website gives access to a vast amount of data on Earth orientation, including the various Bulletins.

Address: International Earth Rotation Service, 61 avenue de l'Observatoire, F-75014 Paris, France.
Website: http://hpiers.obspm.fr

International Astronomical Union (IAU)

The IAU was founded in 1919 to promote and coordinate worldwide co-operation in astronomy. It no longer plays a central role in timekeeping but still maintains a Time Commission which supervises the work of the IERS.

Address: International Astronomical Union, 98bis Bd. Arago, F-75014 Paris, France.
Website: `http://www.iau.org`

International Telecommunication Union (ITU)

The ITU is an international organisation within which governments and the private sector coordinate global telecom networks and services. It is responsible for the international agreements that define UTC in terms of TAI.

Address: International Telecommunication Union, Place des Nations, CH-1211 Geneva 20, Switzerland.
Website: `http://www.itu.int`

National organisations

Many countries have national timing laboratories: the following are just a few that have been mentioned in this book, including all those operating primary frequency standards which contribute to TAI.

Communications Research Laboratory (CRL), Japan

Founded as the Radio Research Laboratory in 1952 but with roots reaching back to 1896, CRL operates a new optically pumped primary standard, CRL-O1, based on the design of NIST-7 and controls the time signals emitted from radio stations JG2AS and JJY.

Address: Communications Research Laboratory, 4-2-1, Nukuikita-machi, Koganei-shi, Tokyo 184, Japan.
Website: `http://www.crl.go.jp`

Institute of Metrology for Time and Space (IMVP), Russia

IMVP operates the caesium beam standard MCsR-102 and a series of time and frequency radio stations including RBU and RWM.

Address: IMVP, GP "VNIIFTRI", Mendeleevo, Moscow Region, 141570, Russia.

Institute for National Measurement Standards (INMS), Canada

The INMS, a laboratory of the National Research Council of Canada, operates two caesium beam standards known as CsVI-A and CsVI-C. It also controls the radio station CHU.

Address: Institute for National Measurement Standards, Montreal Road, Building M-36, Ottawa, Canada, K1A 0R6.
Website: `http://www.nrc.ca/inms/inmse.html`

Laboratoire Primaire du Temps et des Fréquences (LPTF), France

One of the five national laboratories of the Bureau National de Métrologie (BNM), LPTF is responsible for the French time and frequency standards. It operates the world's first caesium fountain, LPTF-FO1, which is the most accurate primary standard to date.

Address: BNM-LPTF, Observatoire de Paris, 61 avenue de l'Observatoire, F-75014 Paris, France.
Website: `http://opdaf1.obspm.fr`

National Institute of Standards and Technology (NIST), USA

Founded as the National Bureau of Standards (NBS) in 1901, NIST hosts the primary standards NIST-7 and NIST-F1 and controls radio stations WWV, WWVH and WWVB. Much of the pioneering research into atomic clocks was carried out at NBS and the first atomic clock, based on a transition in ammonia, was constructed there in 1948.

Address: National Institute of Standards and Technology, 325 Broadway, Boulder, CO 80303-3328, USA.
Website: `http://www.bldrdoc.gov/timefreq`

National Physical Laboratory (NPL), UK

NPL is the UK national standards laboratory, where the world's first operational caesium clock was constructed in 1955. It controls the radio station MSF.

Address: National Physical Laboratory, Teddington, Middlesex TW11 0LW, UK.
Website: http://www.npl.co.uk

National Research Laboratory of Metrology (NRLM), Japan

Now part of the Agency of Industrial Science and Technology, NRLM has been the national standards laboratory of Japan since 1903. It operates NRLM-4, an optically pumped caesium beam standard.

Address: National Research Laboratory of Metrology, 1-1-4, Umezono, Tsukuba, Ibaraki 305-8563, Japan.
Website: http://www.aist.go.jp/NRLM

Physikalisch-Technische Bundesanstalt (PTB), Germany

PTB is Germany's national standards laboratory. In the 1930s PTB scientists were among the first to discover seasonal changes in the Earth's rotation and they now operate the world's most reliable primary frequency standards. PTB also runs radio station DCF-77.

Address: Physikalisch-Technische Bundesanstalt, Postfach 3345, D-38023 Braunschweig, Germany.
Website: http://www.ptb.de

United States Naval Observatory (USNO), USA

USNO plays a leading role in both astronomical and atomic timekeeping, and operates the world's largest ensemble of atomic clocks. In the 1950s USNO astronomers collaborated with NPL scientists to calibrate the atomic second in terms of Ephemeris Time.

Address: United States Naval Observatory, 3450 Massachusetts Avenue NW, Washington, DC 20392-5420, USA.
Website: http://tycho.usno.navy.mil

Other websites about timekeeping

Science Museum (http://www.nmsi.ac.uk/collections/
/exhiblets/atomclock/start.htm)

Properly known as the National Museum of Science and Industry, the London Science Museum is the final resting place of the first operational caesium clock designed by Essen and Parry at NPL.

Royal Observatory Greenwich (http://www.rog.nmm.ac.uk)

The Royal Greenwich Observatory is no more but its legacy is preserved at the old Royal Observatory in Greenwich which is part of the National Maritime Museum, along with much of interest to do with timekeeping.

Long Now Foundation (http://www.longnow.org)

Information about the project to build a monumental clock to last 10 000 years.

GPS (http://www.laafb.af.mil/SMC/CZ/homepage)

Control centre for the US satellite navigation system.

GLONASS (http://mx.iki.rssi.ru/SFCSIC/
/english.html)

Control centre for the Russian satellite navigation system.

Time around the world (http://www.timeanddate.com and
http://www.worldtime.com)

Two sites which can tell you the local time for any place in the world.

Horology (http://www.horology.com)

A guide to many web-based resources about timekeeping.

Glossary of Abbreviations

Timekeeping has spawned its own lexicon of abbreviations and acronyms. The following are the most common examples used in this book.

BIH   Bureau International de l'Heure, the organisation formerly responsible for coodinating world time scales, now superseded by BIPM and IERS.

BIPM   Bureau International des Poids et Mesures, the organisation responsible for international standards of measurement. The Time Section disseminates TAI and UTC.

CGPM   Conférence Générale des Poids et Mesures, the intergovernmental body that oversees the work of CIPM and BIPM.

CIPM   Comité International des Poids et Mesures, the committee of experts that supervises BIPM.

DCF77   A standard time and frequency service broadcast from Germany.

EAL   Echelle Atomique Libre (Free Atomic Time), a free-running time scale formed by averaging the readings of numerous atomic clocks worldwide.

ET   Ephemeris Time, a time scale based on the movements of bodies in the Solar System.

GLONASS   Global Navigation Satellite System, a Russian network of navigational satellites.

GMAT   Greenwich Mean Astronomical Time, a former time scale based on GMT with the date changing at noon rather than midnight.

GMT   Greenwich Mean Time, a former time scale for civil
        timekeeping based on mean solar time at Greenwich,
        now superseded by UTC.

GPS   Global Positioning System, a US network of naviga-
        tional satellites.

IAU   International Astronomical Union, the organisation
        formerly responsible for world timekeeping through
        BIH.

IERS   International Earth Rotation Service, the organisa-
        tion that monitors the Earth's rotation and decides
        when leap seconds are required.

ITU   International Telecommunication Union, the organ-
        isation responsible for the agreements that define
        UTC in terms of TAI.

LPTF   Laboratoire Primaire du Temps et des Fréquences,
        the French time standards laboratory.

MSF   A standard time and frequency service broadcast
        from the UK.

NBS   National Bureau of Standards (US), now renamed as
        NIST.

NIST   National Institute of Standards and Technology, the
        US standards laboratory.

NPL   National Physical Laboratory, the UK standards
        laboratory.

OP   Observatoire de Paris, the national observatory of
        France and the home of LPTF and IERS.

PTB   Physikalisch-Technische Bundesanstalt, the German
        standards laboratory.

PZT   Photographic Zenith Tube, a specialised telescope
        for timing star transits.

RGO   Royal Greenwich Observatory, the former national
        observatory of England.

SI	Système International des Unités, the international system of units.
TAI	Temps Atomique International (International Atomic Time), a time scale formed by calibrating EAL against a small number of primary caesium standards that generate SI seconds.
USNO	United States Naval Observatory, the US national observatory.
UT	Universal Time, originally the scientific name for Greenwich Mean Time but now redefined as UT0, UT1, UT2 and UTC.
UT0	UT determined by astronomical observations at a given observatory.
UT1	UT0 corrected for polar motion.
UT2	UT1 corrected for seasonal variations.
UTC	Coordinated Universal Time, a time scale formed from TAI but kept close to UT1 by the inclusion of leap seconds. The basis of all civil timekeeping.
VLBI	Very Long Baseline Interferometry, a method for determining UT1 and also used for research in radio astronomy.
WWV, WWVB, WWVH	Standard time and frequency services broadcast from the US.

Index